Synthesis Lectures on Intelligent Technologies

Series Editor

Janusz Kacprzyk, Systems Research Institute, Polish Academy of Science, Warsaw, Poland

Synthesis Lectures on Intelligent Technologies provides highly interdisciplinary research with the potential to change the fundamental principles of our society. It covers applications such as Intelligent Transportation, Humanoids, Self-Driving Cars, IoT, Ambient Intelligence, Smart Cities, Human-computer Interaction, Computational Intelligence, Industry 4.0, Medical Robotics, or Data Science. Synthesis Lectures on Intelligent Technologies brings together up-to-date resources from trusted authors working around the world in all aspects of Intelligent Systems.

Mohammad Azad · Igor Chikalov ·
Shahid Hussain · Mikhail Moshkov ·
Beata Zielosko

Decision Trees with Hypotheses

Springer

Mohammad Azad
Department of Computer Science
College of Computer and Information Sciences
Jouf University
Sakakah, Saudi Arabia

Shahid Hussain
Department of Computer Science
School of Mathematics and Computer Science
Institute of Business Administration
Karachi, Pakistan

Beata Zielosko
Faculty of Science and Technology
Institute of Computer Science
University of Silesia in Katowice
Sosnowiec, Poland

Igor Chikalov
Intel Corporation
Chandler, AZ, USA

Mikhail Moshkov
Computer, Electrical and Mathematical
Sciences and Engineering Division
Computational Bioscience Research Center
King Abdullah University of Science
and Technology
Thuwal, Saudi Arabia

ISSN 2731-6912 ISSN 2731-6920 (electronic)
Synthesis Lectures on Intelligent Technologies
ISBN 978-3-031-08584-0 ISBN 978-3-031-08585-7 (eBook)
https://doi.org/10.1007/978-3-031-08585-7

This Springer imprint is published by the registered company Springer Nature Switzerland AG
The registered company address is: Gewerbestrasse 11, 6330 Cham, Switzerland

To our families

Preface

Decision trees are widely used in many areas of computer science and related fields as classifiers, as a means for knowledge representation, and as algorithms to solve various problems. They are studied, in particular, in test theory, rough set theory, and exact learning. These theories are closely related: attributes from rough set theory and test theory correspond to membership queries from exact learning. Exact learning studies additionally the so-called equivalence queries.

In this book, we added to the model considered in test theory and rough set theory the notion of a hypothesis that allowed us to use an analog of equivalence queries and studied decision trees using various combinations of attributes, hypotheses, and proper hypotheses (an analog of proper equivalence queries).

The two main goals of this book are (i) to create tools for the experimental and theoretical study of decision trees with hypotheses and (ii) to compare these decision trees with conventional decision trees that use only queries, each of which is based on one attribute.

We got both experimental and theoretical results showing that the decision trees with hypotheses can have less complexity than the conventional decision trees. These results open up some prospects for using decision trees with hypotheses as a means for knowledge representation and as algorithms for computation of the Boolean functions.

The obtained theoretical results and tools designed for the study of decision trees with hypotheses can be helpful for researchers who use decision trees and rules in data analysis. This book can also be used for the creation of courses for graduate students.

Sakaka, Saudi Arabia — Mohammad Azad
Chandler, USA — Igor Chikalov
Karachi, Pakistan — Shahid Hussain
Thuwal, Saudi Arabia — Mikhail Moshkov
Sosnowiec, Poland — Beata Zielosko
March 2022

Acknowledgements We are greatly indebted to King Abdullah University of Science and Technology for its immense support.

We are thankful to Prof. Andrzej Skowron and to Alexander Knyazev for stimulating discussions.

We extend an expression of gratitude to Prof. Janusz Kacprzyk, Dr. Thomas Ditzinger, and the Series Synthesis Lectures on Intelligent Technologies staff at Springer for their support in making this book possible.

Contents

Introduction

1

Decision trees are widely used in many areas of computer science and related fields as a means for knowledge representation, as classifiers, and as algorithms to solve various problems of combinatorial optimization, computational geometry, etc. [13, 23, 31]. They are studied, in particular, in test theory initiated by Chegis and Yablonskii [14, 33], rough set theory initiated by Pawlak [28–30], and exact learning initiated by Angluin [2–4].

These theories are closely related: attributes from rough set theory and test theory correspond to membership queries from exact learning. Exact learning studies additionally the so-called equivalence queries. The notion of "minimally adequate teacher" that allows both membership and equivalence queries was discussed by Angluin in [2]. Relations between exact learning and PAC learning proposed by Valiant [32] were considered in [3]. Hegedüs in [16] generalized some bounds from [20] obtained in the frameworks of test theory to the case of exact learning with membership and equivalence queries. Similar results were obtained independently and in the other way by Hellerstein et al. [17].

In [22], we moved in the opposite direction: we added to the model considered in test theory and rough set theory the notion of a hypothesis that allowed us to use an analog of equivalence queries, and studied decision trees using various combinations of attributes, hypotheses, and proper hypotheses (an analog of proper equivalence queries) over infinite systems of binary attributes. The paper [22] does not contain proofs. The proofs and some additional results were given in [24, 26]. The translation of the obtained results to the language of exact learning was done in [25]. This new line of research has been continued in the papers [5–12] devoted to the study of optimal decision trees with hypotheses, decision trees constructed by greedy algorithms, and decision rules derived from such trees. Together with [22, 24–26], these papers form the basis of the present book.

The two main goals of this book are (i) to create tools for the experimental and theoretical study of decision trees with hypotheses and (ii) to compare these decision trees with conventional decision trees that use only queries, each of which is based on one attribute. We got both experimental and theoretical results showing that the decision trees with hypotheses

M. Azad et al., *Decision Trees with Hypotheses*, Synthesis Lectures on Intelligent Technologies, https://doi.org/10.1007/978-3-031-08585-7_1

can have less complexity than the conventional decision trees. These results open up some prospects for using decision trees with hypotheses as a means for knowledge representation and as algorithms for the computation of Boolean functions.

This book consists of two parts. The first part, "Decision Tables," corresponds to the case when we have only one problem represented by a decision table. The second part, "Infinite Binary Information Systems and Infinite Families of Concepts," corresponds to the case when we have an infinite set of attributes represented as an infinite binary information system and any finite subset of this set of attributes forms a problem. The words "Infinite Families of Concepts" are related to the translation of the obtained results into the language of exact learning.

The rest of the chapter is organized as follows. Sections 1.1 and 1.2 summarize Parts I and II of the present book. In Sect. 1.3, we discuss prospects of using decision trees with hypotheses, and in Sect. 1.4—the use of the book.

1.1 Part I. Decision Tables

In this part, which is based on [5–12], we introduce and study the notion of decision tree with hypotheses for a decision table.

Let T be a decision table with n conditional attributes $f_1, \ldots, f_n$ having values from the set $\omega = \{0, 1, 2, \ldots\}$ in which rows are pairwise different and each row is labeled with a decision from ω. For a given row of T, we should recognize the decision attached to this row. To this end, we can use decision trees based on two types of queries. We can ask about the value of an attribute $f_i \in \{f_1, \ldots, f_n\}$ on the given row. We will obtain an answer of the kind $f_i = \delta$, where δ is the number in the intersection of the given row and the column f_i. We can also ask if a hypothesis $f_1 = \delta_1, \ldots, f_n = \delta_n$ is true, where $\delta_1, \ldots, \delta_n$ are numbers from the columns $f_1, \ldots, f_n$, respectively. Either this hypothesis will be confirmed, or we obtain a counterexample in the form $f_i = \sigma$, where $f_i \in \{f_1, \ldots, f_n\}$ and σ is a number from the column f_i different from δ_i. The considered hypothesis is called proper if $(\delta_1, \ldots, \delta_n)$ is a row of the table T.

We study the following five types of decision trees:

1. Decision trees that use only attributes.
2. Decision trees that use only hypotheses.
3. Decision trees that use both attributes and hypotheses.
4. Decision trees that use only proper hypotheses.
5. Decision trees that use both attributes and proper hypotheses.

We consider four cost functions that characterize the time and space complexity of decision trees. We consider the depth of a decision tree as its time complexity, which is equal to the maximum number of queries in a path from the root to a terminal node of the tree. As the space complexity of a decision tree, we consider the number of its realizable relative to T nodes. A node is called realizable relative to the decision table T if, for a row of T and some choice of counterexamples, the computation in the tree will pass through this node. In addition to the number of realizable nodes, we also consider the number of realizable terminal nodes and the number of working (nonterminal) nodes. Note that in the considered trees, all working nodes are realizable.

To investigate decision trees with hypotheses, we created algorithms that allow us (i) to find the complexity of optimal decision trees and to construct such trees for medium-sized decision tables using dynamic programming approach, (ii) to construct decision trees using various greedy heuristics, and (iii) to derive decision rules from the constructed decision trees and study their length and coverage. Implementations of these algorithms were added to the Dagger software system developed in KAUST and intended mainly for the study of decision and inhibitory trees and rules [1].

We study not only decision trees but also decision rules since they can be considered one of the simplest and most understandable models for knowledge representation. Deriving decision rules from decision trees is a well-known approach. For a decision tree with hypotheses for some rows of the considered decision table, it can be more than one derived decision rule that covers the row. In this case, for each row, we choose the best rule.

We made computer experiments on 10 decision tables from the UCI ML Repository [15] and 100 randomly generated Boolean functions with n variables, where $n = 3, \ldots, 6$. The table representation of a Boolean function can be interpreted as a decision table. Decision trees for this decision table compute the considered Boolean function.

From the obtained experimental results it follows that (i) the optimal decision trees of types 3 and 5 are, generally, slightly better than the optimal decision trees of type 1, (ii) using hypotheses we can, generally, decrease the depth of the decision trees constructed by greedy algorithms but not the number of realizable nodes, and (iii) the decision rules derived from the constructed decision trees with hypotheses in many cases are better from the point of view of the length and coverage than the rules derived from the constructed conventional decision trees.

The study of experimental results allowed us to formulate interesting hypotheses that were later proved. In particular, we confirmed that, for any decision table, we can construct a decision tree with the minimum number of realizable terminal nodes using only attributes.

Note that the created tools can be used for the study of exact learning problems. An exact learning problem is defined by a domain X and a concept class B [4]. The domain X is a nonempty finite set. A concept is any subset of X, and a concept class B is a nonempty set of concepts. For a given (but hidden) concept $c \in B$, we should recognize it using two types of queries. A membership query: for an element $x \in X$, we ask if $x \in c$. The answer is either $x \in c$ or $x \notin c$. An equivalence query: for a concept $c' \subseteq X$, we ask if $c' = c$. The answer is either yes or no. In the latter case, we receive a counterexample $x \in (c' \setminus c) \cup (c \setminus c')$, which is chosen nondeterministically. The considered equivalence query is called proper if $c' \in B$.

Let $X = \{x_1, \ldots, x_n\}$ and $B = \{c_1, \ldots, c_m\}$. We can correspond to the exact learning problem (X, B) a decision table $T(X, B)$ with n columns and m rows. Columns are labeled with "attributes" $x_1, \ldots, x_n$ and rows are labeled with "decisions" $c_1, \ldots, c_m$, respectively. This table is filled with numbers from the set $\{0, 1\}$. In the intersection with the column x_i, the row c_j has 1 if and only if $x_i \in c_j$. Note that a similar table representation for exact learning problems was considered in [4]. Instead of studying decision trees based on membership and equivalence queries for the problem (X, B), we can study the decision trees based on attributes and hypotheses for the decision table $T(X, B)$.

As illustrative examples, we consider the problem of recognition of monotone Boolean functions with n variables, $n = 2, \ldots, 4$, and the problem of sorting n pairwise different elements from linearly ordered set, $n = 3, \ldots, 6$, which can be reformulated as exact learning problems.

1.2 Part II. Infinite Binary Information Systems and Infinite Families of Concepts

In this part (see also [22, 24–26]), based on the results of exact learning [3, 4, 18, 19] and test theory and rough set theory [20, 21, 23], we investigate infinite binary information systems (similar to the information systems introduced by Pawlak [27]).

Each such system consists of an infinite set of elements A and an infinite set F of functions (attributes) from A to $\{0, 1\}$. We define the notion of a testing problem described by a finite number of attributes $f_1, \ldots, f_n$ from F: for a given element $a \in A$, we should recognize the tuple $(f_1(a), \ldots, f_n(a))$. To this end, we can use decision trees based on two types of queries. We can ask about the value of an attribute $f_i \in \{f_1, \ldots, f_n\}$. We will obtain an answer of the kind $f_i(a) = \delta$, where $\delta \in \{0, 1\}$. We can also ask if a hypothesis $f_1(a) = \delta_1, \ldots, f_n(a) = \delta_n$ is true, where $\delta_1, \ldots, \delta_n \in \{0, 1\}$. This hypothesis will either be confirmed or we will obtain a counterexample $f_i(a) = \neg\delta_i$ chosen nondeterministically. The considered hypothesis is called proper if there exists an element $b \in A$ such that $f_1(b) = \delta_1, \ldots, f_n(b) = \delta_n$. As time complexity of a decision tree, we consider its depth equal to the maximum number of queries in a path from the root to a terminal node.

For an arbitrary infinite binary information system, we study five functions of Shannon type, which characterize the dependence in the worst case of the minimum depth of a decision tree solving a testing problem on the number of attributes in the problem description. These five functions correspond to the following five types of decision trees: (1) using attributes, (2) using hypotheses, (3) using attributes and hypotheses, (4) using proper hypotheses, and (5) using attributes and proper hypotheses.

We prove that the first function has two possible types of behavior: logarithmic and linear. Each of the remaining four functions has three possible types of behavior: constant, logarithmic, and linear. We also study the joint behavior of all five functions and describe seven complexity classes of infinite binary information systems.

From the obtained results, it follows, in particular, that there are infinite binary information systems for which the first function corresponding to the conventional decision trees has linear growth but the rest four functions corresponding to the decision trees with hypotheses are bounded from above by a constant.

To translate the results obtained for infinite binary information systems into exact learning, we introduce the notion of an infinite family of concepts consisting of an infinite set of elements U and an infinite set C of subsets of U called concepts. Each nonempty finite subset X of the set U describes a problem of exact learning with the domain X and the concept class $\{c \cap X : c \in C\}$. We study the following modification of this problem: for a given (but hidden) concept $c \in C$, we should recognize the concept $c \cap X$ using the same queries as for the initial problem. Both the problem and its modification have the same sets of decision trees solving them.

For each infinite family of concepts, we study five functions of Shannon type, which characterize the dependence in the worst case of the minimum depth of a decision tree solving an exact learning problem on the number of elements in the problem description. These five functions correspond to the following five cases:

1. Only membership queries are used in decision trees.
2. Only equivalence queries are used in decision trees.
3. Both membership and equivalence queries are used in decision trees.
4. Only proper equivalence queries are used in decision trees.
5. Both membership and proper equivalence queries are used in decision trees.

As in the case of testing problems, the first function has two possible types of behavior—logarithmic and linear, and each of the remaining functions has three possible types of behavior—constant, logarithmic, and linear. We also consider the joint behavior of these five functions and distinguish seven complexity classes of infinite families of concepts.

The translation of the results obtained for decision trees with hypotheses into exact learning will be helpful for researchers in this area: the duality of the two directions is simple (attributes and elements from test theory and rough set theory correspond to elements and concepts from exact learning, respectively) but working out the details takes some effort.

1.3 Prospects of Using Decision Trees with Hypotheses

Undoubtedly, decision trees with hypotheses are an interesting model for theoretical research. In this book, we do not set ourselves the goal of investigating the application of such trees, we only want to outline these possible applications.

The practical use of decision trees with hypotheses is limited because we usually cannot effectively implement hypothesis-based queries: straightforward methods require sequential computation of values of all attributes. Generally, we cannot use such trees as classifiers and as algorithms. However, there is a chance that we will be able to use these trees as a means for knowledge representation and as algorithms for the computation of Boolean functions. In the former case, we do not need to implement the queries. In the latter case, we already know the values of all attributes (variables). As a corroborating example, we consider the problem of computation of the conjunction $x_1 \wedge \cdots \wedge x_{64}$.

The minimum depth of a conventional decision tree solving this problem using the attributes $x_1, \ldots, x_{64}$ is 64. The minimum number of realizable nodes is 129, the minimum number of working nodes is 64, and the minimum number of realizable terminal nodes is 65. Among decision rules derived from each conventional decision tree, there are 64 pairwise different rules $r_1, \ldots, r_{64}$ such that, for $i = 1, \ldots, 64$, the length of the rule r_i is at least i, and the coverage of this rule is at most 2^{64-i}. The number of derived rules is at least 65. One of the conventional decision trees for the computation of the function $x_1 \wedge \cdots \wedge x_{64}$ is depicted in Fig. 1.1. For technical reasons, it will be convenient for us to represent answers to queries in the form of systems of equations of the kind "attribute = value".

The minimum depth of a decision tree solving this problem using proper hypotheses is 1: it is enough to ask only about the hypothesis $x_1 = 1, \ldots, x_{64} = 1$ denoted by H. If it is true, then the considered conjunction equals 1. Otherwise, it is 0. The obtained decision tree contains one working node and 65 realizable terminal nodes, altogether 66 realizable nodes. Among 65 decision rules derived from this tree, one has the length 64 and the coverage 1. For each of the remaining rules, the length is 1 and the coverage is 2^{63}. The considered decision tree is depicted in Fig. 1.2.

Fig. 1.1 Conventional decision tree computing $x_1 \wedge \cdots \wedge x_{64}$

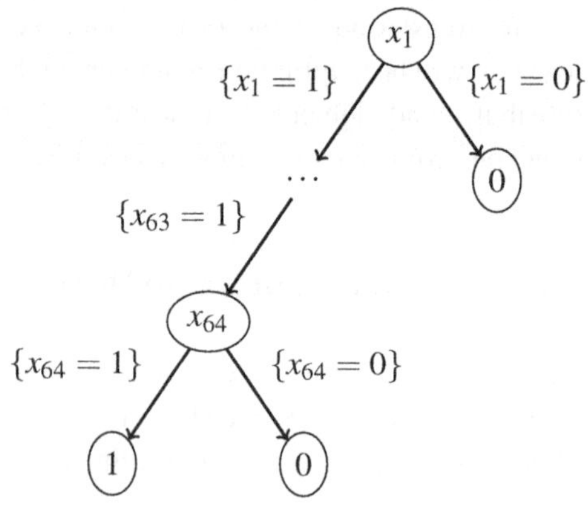

Fig. 1.2 Decision tree with hypotheses computing $x_1 \wedge \cdots \wedge x_{64}$

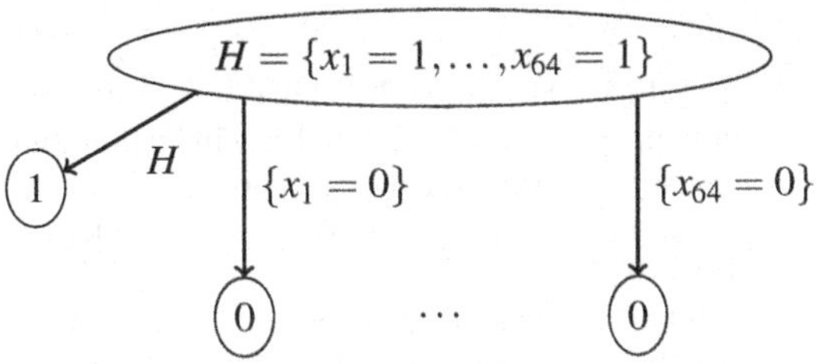

We see that, for the conjunction of 64 variables, both the proposed decision tree with hypotheses and the decision rules derived from it are better from the point of view of the considered parameters than the conventional decision trees and decision rules derived from them.

Various experimental results in Part I show that the decision trees with hypotheses can have less complexity than the conventional decision trees and that the decision rules derived from the former trees have better length and coverage than the decision rules derived from the latter trees.

We now describe a simple way to simulate a query based on the hypothesis H. We compare the tuple $(x_1, \ldots, x_{64})$ and the tuple $(1, \ldots, 1)$. This operation usually requires one clock cycle. If the tuples are equal, then the hypothesis H is true and the conjunction equals 1. Otherwise, the hypothesis H is false and the conjunction equals 0. In this particular case, we do not need to return a counterexample.

Details of the simulation of arbitrary queries based on hypotheses for the computation of Boolean functions, including the process of returning the counterexamples as well as examples of decision trees with hypotheses for some classes of Boolean functions can be found in Appendix A.

1.4 Use of Book

The obtained theoretical results and tools designed for the study of decision trees with hypotheses can be helpful for researchers who use decision trees and rules in data analysis. In particular, the results presented in this book can be useful in rough set theory, test theory, and exact learning. This book can also be used for the creation of courses for graduate students.

References

1. AbouEisha, H., Amin, T., Chikalov, I., Hussain, S., Moshkov, M.: Extensions of Dynamic Programming for Combinatorial Optimization and Data Mining. Intelligent Systems Reference Library, vol. 146. Springer (2019)
2. Angluin, D.: Learning regular sets from queries and counterexamples. Inf. Comput. **75**(2), 87–106 (1987)
3. Angluin, D.: Queries and concept learning. Mach. Learn. **2**(4), 319–342 (1988)
4. Angluin, D.: Queries revisited. Theor. Comput. Sci. **313**(2), 175–194 (2004)
5. Azad, M., Chikalov, I., Hussain, S., Moshkov, M.: Entropy-based greedy algorithm for decision trees using hypotheses. Entropy **23**(7), 808 (2021). https://doi.org/10.3390/e23070808
6. Azad, M., Chikalov, I., Hussain, S., Moshkov, M.: Minimizing depth of decision trees with hypotheses. In: Ramanna, S., Cornelis, C., Ciucci, D.(eds.) Rough Sets—International Joint Conference, IJCRS 2021, Bratislava, Slovakia, 19–24 Sept 2021. Lecture Notes in Computer Science, vol. 12872, pp. 123–133. Springer (2021)
7. Azad, M., Chikalov, I., Hussain, S., Moshkov, M.: Minimizing number of nodes in decision trees with hypotheses. In: Watrobski, J., Salabun, W., Toro, C., Zanni-Merk, C., Howlett, R.J., Jain, L.C. (eds.) 25th International Conference on Knowledge-Based and Intelligent Information & Engineering Systems, KES 2021, 8–10 Sept 2021, Szczecin, Poland. Procedia Computer Science, vol. 192, pp. 232–240. Elsevier (2021). https://doi.org/10.1016/j.procs.2021.08.024
8. Azad, M., Chikalov, I., Hussain, S., Moshkov, M.: Optimization of decision trees with hypotheses for knowledge representation. Electronics **10**(13), 1580 (2021). https://doi.org/10.3390/electronics10131580
9. Azad, M., Chikalov, I., Hussain, S., Moshkov, M.: Sorting by decision trees with hypotheses (extended abstract). In: Schlingloff, H., Vogel, T. (eds.) 29th International Workshop on Concurrency, Specification and Programming, CS&P 2021, Berlin, Germany, 27–28 Sept 2021. CEUR Workshop Proceedings, vol. 2951, pp. 126–130. CEUR-WS.org (2021). http://ceur-ws.org/Vol-2951/paper1.pdf
10. Azad, M., Chikalov, I., Hussain, S., Moshkov, M., Zielosko, B.: Decision rules derived from optimal decision trees with hypotheses. Entropy **23**(12), 1641 (2021). https://doi.org/10.3390/e23121641
11. Azad, M., Chikalov, I., Hussain, S., Moshkov, M., Zielosko, B.: Decision trees with hypotheses for recognition of monotone Boolean functions and for sorting. arXiv:2203.08894 [cs.CC] (2022). https://doi.org/10.48550/arXiv.2203.08894
12. Azad, M., Chikalov, I., Hussain, S., Moshkov, M., Zielosko, B.: Greedy algorithms for decision trees with hypotheses. arXiv:2203.08848 [cs.CC] (2022). https://doi.org/10.48550/arXiv.2203.08848
13. Breiman, L., Friedman, J.H., Olshen, R.A., Stone, C.J.: Classification and Regression Trees. Chapman and Hall/CRC (1984)
14. Chegis, I.A., Yablonskii, S.V.: Logical methods of control of work of electric schemes. Trudy Mat. Inst. Steklov **51**, 270–360 (1958). (in Russian)
15. Dua, D., Graff, C.: UCI Machine Learning Repository. University of California, Irvine, School of Information and Computer Sciences (2017). http://archive.ics.uci.edu/ml
16. Hegedüs, T.: Generalized teaching dimensions and the query complexity of learning. In: Maass, W. (ed.) Eight Annual Conference on Computational Learning Theory, COLT 1995, Santa Cruz, CA, USA, 5–8 July 1995, pp. 108–117. ACM (1995)
17. Hellerstein, L., Pillaipakkamnatt, K., Raghavan, V., Wilkins, D.: How many queries are needed to learn? J. ACM **43**(5), 840–862 (1996)

18. Littlestone, N.: Learning quickly when irrelevant attributes abound: a new linear-threshold algorithm. Mach. Learn. **2**(4), 285–318 (1988)
19. Maass, W., Turán, G.: Lower bound methods and separation results for on-line learning models. Mach. Learn. **9**, 107–145 (1992)
20. Moshkov, M.: Conditional tests. In: Yablonskii, S.V. (ed.) Problemy Kibernetiki, vol. 40, pp. 131–170. Nauka Publishers, Moscow (1983). (in Russian)
21. Moshkov, M.: On depth of conditional tests for tables from closed classes. In: Markov, A.A. (ed.) Combinatorial-Algebraic and Probabilistic Methods of Discrete Analysis, pp. 78–86. Gorky University Press, Gorky (1989). (in Russian)
22. Moshkov, M.: Test theory and problems of machine learning. In: International School-Seminar on Discrete Mathematics and Mathematical Cybernetics, Ratmino, Russia, 31 May–3 June 2001, pp. 6–10. MAX Press, Moscow (2001)
23. Moshkov, M.: Time complexity of decision trees. In: Peters, J.F., Skowron, A. (eds.) Trans. Rough Sets III. Lecture Notes in Computer Science, vol. 3400, pp. 244–459. Springer (2005)
24. Moshkov, M.: Exact learning and test theory. arXiv:2201.04506 [cs.CC] (2022). https://doi.org/10.48550/arXiv.2201.04506
25. Moshkov, M.: Exact learning for infinite families of concepts. arXiv:2201.08225 [cs.AI] (2022). https://doi.org/10.48550/arXiv.2201.08225
26. Moshkov, M.: On the depth of decision trees with hypotheses. Entropy **24**(1), 116 (2022). https://doi.org/10.3390/e24010116
27. Pawlak, Z.: Information systems theoretical foundations. Inf. Syst. **6**(3), 205–218 (1981)
28. Pawlak, Z.: Rough sets. Int. J. Parallel Program. **11**(5), 341–356 (1982)
29. Pawlak, Z.: Rough Sets—Theoretical Aspects of Reasoning About Data. Theory and Decision Library: Series D, vol. 9. Kluwer (1991)
30. Pawlak, Z., Skowron, A.: Rudiments of rough sets. Inf. Sci. **177**(1), 3–27 (2007)
31. Rokach, L., Maimon, O.: Data Mining with Decision Trees—Theory and Applications. Series in Machine Perception and Artificial Intelligence, vol. 69. World Scientific (2007)
32. Valiant, L.G.: A theory of the learnable. Commun. ACM **27**(11), 1134–1142 (1984)
33. Yablonskii, S.V., Chegis, I.A.: On tests for electric circuits. Usp. Mat. Nauk **10**(4), 182–184 (1955). (in Russian)

Part I
Decision Tables

In this part, we study decision trees with hypotheses for decision tables. We consider dynamic programming algorithms for decision tree optimization and greedy algorithms for decision tree construction. We discuss results of computational experiments on decision tables from the UCI Machine Learning Repository, on randomly generated Boolean functions, on decision tables corresponding to the problem of recognition of monotone Boolean functions, and on decision tables corresponding to the problem of sorting. We study not only time and space complexity of constructed decision trees but also the length and coverage of decision rules derived from these trees.

In Chap. 2, we consider main notions for Part I: decision tables and uncertainty measures, decision trees, and decision rules derived from decision trees.

In Chap. 3, we present dynamic programming algorithms for the minimization of the depth and number of nodes in decision trees and discuss results of computer experiments on eight data sets from the UCI Machine Learning Repository and randomly generated Boolean functions.

In Chap. 4, we design dynamic programming algorithms for the construction of optimal decision trees relative to two cost functions, construct such trees for eight datasets from the UCI Machine Learning Repository and randomly generated Boolean functions, and compare the length and coverage of decision rules derived from the constructed decision trees.

In Chap. 5, we present greedy algorithms based on diverse uncertainty measures for construction of decision trees with hypotheses and discuss results of computer experiments on 10 datasets from the UCI Machine Learning Repository and randomly generated Boolean functions. We also study the length and coverage of decision rules derived from the decision trees constructed by greedy algorithms.

In Chap. 6, we study the problem of recognition of monotone Boolean functions with n variables, $n = 2, \ldots, 4$, and the problem of sorting n pairwise different elements from linearly ordered set, $n = 3, \ldots, 6$. For each of these problems, we compare the complexity of five types of optimal (relative to the depth or the number of realizable nodes) decision trees. We also study the complexity of decision trees constructed by greedy algorithms and analyze the length of decision rules derived from these trees.

Main Notions

2

In this chapter, we define main notions for Part I. We begin with the consideration of conventional decision tables with one-valued decisions and uncertainty measures for these tables that are used by greedy algorithms for decision tree construction.

After that, we define five types of decision trees based on different combinations of attributes, hypotheses, and proper hypotheses. We also consider four cost functions that characterize the time and space complexity of decision trees.

At the end, we define a set of decision rules derived from a decision tree and discuss two parameters of this set related to the length and coverage of decision rules.

The definitions discussed in this chapter are close to ones considered in [1–3].

The rest of the chapter is organized as follows. In Sect. 2.1, we consider decision tables and uncertainty measures, in Sect. 2.2—decision trees, and in Sect. 2.3—decision rules derived from decision trees.

2.1 Decision Tables and Uncertainty Measures

A *decision table* is a rectangular table T with $n \geq 1$ columns filled with numbers from the set $\omega = \{0, 1, 2, \ldots\}$ of nonnegative integers. Columns of this table are labeled with the *conditional attributes* $f_1, \ldots, f_n$. Rows of the table are pairwise different. Each row is labeled with a number from ω that is interpreted as a decision. Rows of the table are interpreted as tuples of values of the conditional attributes.

A decision table can be represented by a word over the alphabet $\{0, 1, ;, |\}$ in which numbers from ω are in binary representation (are represented by words over the alphabet $\{0, 1\}$), the symbol ";" is used to separate two numbers from ω, and the symbol "|" is used to separate two rows (we add to each row corresponding decision as the last number in the row). The length of this word will be called the *size* of the decision table.

M. Azad et al., *Decision Trees with Hypotheses*, Synthesis Lectures on Intelligent Technologies, https://doi.org/10.1007/978-3-031-08585-7_2

A decision table T is called *empty* if it has no rows. The table T is called *degenerate* if it is empty or all rows of T are labeled with the same decision.

We denote $F(T) = \{f_1, \ldots, f_n\}$ and denote by $D(T)$ the set of decisions attached to the rows of T. For any conditional attribute $f_i \in F(T)$, we denote by $E(T, f_i)$ the set of values of the attribute f_i in the table T. We denote by $E(T)$ the set of conditional attributes of T for which $|E(T, f_i)| \geq 2$.

A *system of equations over* T is an arbitrary equation system of the kind

$$\{f_{i_1} = \delta_1, \ldots, f_{i_m} = \delta_m\},$$

where $m \in \omega$, $f_{i_1}, \ldots, f_{i_m} \in F(T)$, and $\delta_1 \in E(T, f_{i_1}), \ldots, \delta_m \in E(T, f_{i_m})$ (if $m = 0$, then the considered equation system is empty).

Let T be a nonempty table. A *subtable* of T is a table obtained from T by removal of some rows. We correspond to each equation system S over T a subtable TS of the table T. If the system S is empty, then $TS = T$. Let S be nonempty and $S = \{f_{i_1} = \delta_1, \ldots, f_{i_m} = \delta_m\}$. Then TS is the subtable of the table T containing the rows from T, which in the intersection with the columns $f_{i_1}, \ldots, f_{i_m}$ have numbers $\delta_1, \ldots, \delta_m$, respectively. Such nonempty subtables, including the table T, are called *separable* subtables of T. We denote by $SEP(T)$ the set of separable subtables of the table T.

We denote by $N(T)$ the number of rows in the table T and, for each $t \in \omega$, we denote by $N_t(T)$ the number of rows in the table T that are labeled with the decision t. By $mcd(T)$ we denote the minimum number $t_0 \in \omega$ such that $N_{t_0}(T) = \max\{N_t(T) : t \in \omega\}$. The value $mcd(T)$ is called the *most common decision* for T.

We denote by $\mathscr{T}$ the set of all decision tables. Let $\mathbb{R}$ be the set of real numbers. An *uncertainty measure* is a function $U : \mathscr{T} \to \mathbb{R}$ such that $U(T) \geq 0$ for any $T \in \mathscr{T}$ and $U(T) = 0$ if and only if T is a degenerate table. One can show (see book [1]) that the following functions (we assume that, for any empty table, the value of each of the considered functions is equal to 0) are uncertainty measures:

- Misclassification error $me(T) = N(T) - N_{mcd(T)}(T)$.
- Relative misclassification error $rme(T) = (N(T) - N_{mcd(T)}(T))/N(T)$.
- Entropy $ent(T) = -\sum_{t \in D(T)}(N_t(T)/N(T)) \log_2(N_t(T)/N(T))$.
- Gini index $gini(T) = 1 - \sum_{t \in D(T)}(N_t(T)/N(T))^2$.
- Function R, where $R(T)$ is the number of unordered pairs of rows of T labeled with different decisions (note that $R(T) = N(T)^2 gini(T)/2$).

2.2 Decision Trees

Let T be a nonempty decision table with n conditional attributes $f_1, \ldots, f_n$. We consider the decision trees with two types of *queries*. We can choose an attribute $f_i \in F(T) = \{f_1, \ldots, f_n\}$ and ask about its value. This query has the set of answers $A(f_i) = \{\{f_i =$

$\delta\} : \delta \in E(T, f_i)\}$. We can formulate a *hypothesis* over T in the form of $H = \{f_1 = \delta_1, \ldots, f_n = \delta_n\}$, where $\delta_1 \in E(T, f_1), \ldots, \delta_n \in E(T, f_n)$, and ask about this hypothesis. This query has the set of answers $A(H) = \{H, \{f_1 = \sigma_1\}, \ldots, \{f_n = \sigma_n\} : \sigma_1 \in E(T, f_1) \setminus \{\delta_1\}, \ldots, \sigma_n \in E(T, f_n) \setminus \{\delta_n\}\}$. The answer H means that the hypothesis is true. Other answers are counterexamples. The hypothesis H is called *proper* for T if $(\delta_1, \ldots, \delta_n)$ is a row of the table T.

A *decision tree over* T is a marked finite directed tree with the root in which

- Each terminal node is labeled with a number from the set $D(T) \cup \{0\}$.
- Each node, which is not terminal (such nodes are called *working*), is labeled with an attribute from the set $F(T)$ or with a hypothesis over T.
- If a working node is labeled with an attribute f_i from $F(T)$, then, for each answer from the set $A(f_i)$, there is exactly one edge labeled with this answer, which leave this node and there are no any other edges leaving this node.
- If a working node is labeled with a hypothesis $H = \{f_1 = \delta_1, \ldots, f_n = \delta_n\}$ over T, then, for each answer from the set $A(H)$, there is exactly one edge labeled with this answer, which leaves this node and there are no any other edges leaving this node.

Let Γ be a decision tree over T and v be a node of Γ. We now define an equation system $S(\Gamma, v)$ over T associated with the node v. We denote by ξ the directed path from the root of Γ to the node v. If there are no working nodes in ξ, then $S(\Gamma, v)$ is the empty system. Otherwise, $S(\Gamma, v)$ is the union of equation systems attached to the edges of the path ξ.

A decision tree Γ over T is called a *decision tree for* T if, for any node v of Γ,

- The node v is terminal if and only if the subtable $TS(\Gamma, v)$ is degenerate.
- If v is a terminal node and the subtable $TS(\Gamma, v)$ is empty, then the node v is labeled with the decision 0.
- If v is a terminal node and the subtable $TS(\Gamma, v)$ is nonempty, then the node v is labeled with the decision attached to all rows of $TS(\Gamma, v)$.

A *complete path* in Γ is an arbitrary directed path from the root to a terminal node in Γ. As the time complexity of a decision tree, we consider its *depth* that is the maximum number of working nodes in a complete path in the tree or, which is the same, the maximum length of a complete path in the tree. We denote by $h(\Gamma)$ the depth of a decision tree Γ.

As the space complexity of the decision tree Γ, we consider the number of its *realizable* relative to T nodes. A node v of Γ is called realizable relative to T if and only if the subtable $TS(\Gamma, v)$ is nonempty. We denote by $L(T, \Gamma)$ the number of nodes in Γ that are realizable relative to T. We also consider two more cost functions relative to the space complexity: $L_t(T, \Gamma)$—the number of terminal nodes in Γ that are realizable relative to T and $L_w(T, \Gamma)$—-the number of working nodes in Γ. Note that all working nodes of Γ are realizable relative to T.

We consider the following five *types* of decision trees:

1. Decision trees that use only attributes.
2. Decision trees that use only hypotheses.
3. Decision trees that use both attributes and hypotheses.
4. Decision trees that use only proper hypotheses.
5. Decision trees that use both attributes and proper hypotheses.

We will use the following notation:

- For $k = 1, \ldots, 5$, $h^{(k)}(T)$ is the minimum depth of a decision tree of type k for T.
- For $k = 1, \ldots, 5$, $L^{(k)}(T)$ is the minimum number of nodes realizable relative to T in a decision tree of type k for T.
- For $k = 1, \ldots, 5$, $L_t^{(k)}(T)$ is the minimum number of terminal nodes realizable relative to T in a decision tree of type k for T.
- For $k = 1, \ldots, 5$, $L_w^{(k)}(T)$ is the minimum number of working nodes in a decision tree of type k for T.

2.3 Decision Rules Derived from Decision Trees

Let Γ be a decision tree for T and ξ be a complete path in Γ such that the terminal node u of ξ is labeled with a decision d and the subtable $T(\xi)$ is nonempty, where $T(\xi) = TS(\Gamma, u)$. We now define a system of equations $S(\xi)$. If there are no working nodes in ξ, then $S(\xi)$ is the empty system. Let us assume now that ξ contains at least one working node. We now transform systems of equations attached to edges leaving working nodes of ξ. If an edge is labeled with an equation system containing exactly one equation, then we will not change this system. Let an edge e leaving a working node v be labeled with an equation system containing more than one equation. Then v is labeled with a hypothesis H and e is labeled with the equation system H. (Note that if such a node exists, then it is the last working node in the complete path ξ.) In this case, we remove from the equation system H attached to e all equations of the kind $f_j = \sigma$ such that $f_j \notin E(TS(\Gamma, v))$. Then $S(\xi)$ is the union of new equation systems attached to the edges of the path ξ. The removed equations were redundant: one can show that $T(\xi) = TS(\xi)$.

We correspond to the complete path ξ the *decision rule*

$$\bigwedge_{f_i = \delta \in S(\xi)} (f_i = \delta) \rightarrow d.$$

We denote this rule by $rule(\xi)$. The number of equations in the equation system $S(\xi)$ is called the *length* of the rule $rule(\xi)$ and is denoted $l(rule(\xi))$. The number of rows in the subtable $TS(\xi)$ is called the *coverage* of the rule $rule(\xi)$ and is denoted $c(rule(\xi))$.

Denote $\Xi(T, \Gamma)$ the set of complete paths ξ in Γ such that the table $T(\xi)$ is nonempty and $Rows(T)$ the set of rows of the decision table T. For a row $r \in Rows(T)$, we denote by $l(r, T, \Gamma)$ the minimum length of a rule $rule(\xi)$ such that $\xi \in \Xi(T, \Gamma)$ and r is a row of the subtable $TS(\xi)$, and we denote by $c(r, T, \Gamma)$ the maximum coverage of a rule $rule(\xi)$ such that $\xi \in \Xi(T, \Gamma)$ and r is a row of the subtable $TS(\xi)$.

We will use the following notation:

$$l(T, \Gamma) = \frac{\sum_{r \in Rows(T)} l(r, T, \Gamma)}{|Rows(T)|},$$

$$c(T, \Gamma) = \frac{\sum_{r \in Rows(T)} c(r, T, \Gamma)}{|Rows(T)|}.$$

References

1. AbouEisha, H., Amin, T., Chikalov, I., Hussain, S., Moshkov, M.: Extensions of Dynamic Programming for Combinatorial Optimization and Data Mining, Intelligent Systems Reference Library, vol. 146. Springer (2019)
2. Azad, M., Chikalov, I., Hussain, S., Moshkov, M.: Optimization of decision trees with hypotheses for knowledge representation. Electronics **10**(13), 1580 (2021). https://doi.org/10.3390/electronics10131580
3. Azad, M., Chikalov, I., Hussain, S., Moshkov, M., Zielosko, B.: Decision rules derived from optimal decision trees with hypotheses. Entropy **23**(12), 1641 (2021). https://doi.org/10.3390/e23121641

Dynamic Programming Algorithms for Minimization of Decision Tree Complexity

3

In this chapter, which is based on two conference papers [3, 4] and the journal publication [5] summarizing them, we study decision trees with hypotheses. Our goal is to understand whether it is possible to reduce the time and space complexity of conventional decision trees if we use additionally hypotheses. Decision trees with less complexity are more understandable and more suitable as a means for knowledge representation.

We study the following five types of decision trees:

1. Decision trees that use only attributes.
2. Decision trees that use only hypotheses.
3. Decision trees that use both attributes and hypotheses.
4. Decision trees that use only proper hypotheses.
5. Decision trees that use both attributes and proper hypotheses.

We consider four cost functions that characterize the time and space complexity of decision trees: the depth, the number of realizable nodes, the number of realizable terminal nodes, and the number of working (nonterminal) nodes.

For each cost function, we propose a dynamic programming algorithm that, for a given decision table and given type of decision trees, finds the minimum cost of a decision tree of the considered type for this table. We can use these algorithms to find the minimum cost of decision trees for medium-sized decision tables. Note that dynamic programming algorithms for the optimization of decision trees of type 1 were studied in [1] for decision tables with one-valued decisions and in [2] for decision tables with many-valued decisions.

For each cost function, we compute the minimum cost of a decision tree for each of the considered five types for eight decision tables from the UCI ML Repository [6]. We do the same for 100 randomly generated Boolean functions with n variables, where $n = 3, \ldots, 6$.

M. Azad et al., *Decision Trees with Hypotheses*, Synthesis Lectures on Intelligent Technologies, https://doi.org/10.1007/978-3-031-08585-7_3

From the obtained experimental results it follows that, generally, the decision trees of types 3 and 5 have less complexity than the decision trees of type 1. Therefore such decision trees can be useful as a means for knowledge representation. Decision trees of types 2 and 4 have, generally, too many nodes.

Based on the experimental results, we formulate and prove the following hypothesis: for any decision table, we can construct a decision tree with the minimum number of realizable terminal nodes using only attributes.

The rest of the chapter is organized as follows. In Sects. 3.1–3.5, we consider dynamic programming algorithms for the decision tree optimization. In Sect. 3.6, we prove the above hypothesis. Section 3.7 contains results of computer experiments and Sect. 3.8—short conclusions.

3.1 Construction of Directed Acyclic Graph $\Delta(T)$

Let T be a nonempty decision table with n conditional attributes $f_1, \dots, f_n$. We now consider an algorithm $\mathscr{A}_{DAG}$ for the construction of a *directed acyclic graph* (DAG) $\Delta(T)$, which will be used for the study of decision trees. Nodes of this graph are some separable subtables of the table T. During each iteration we process one node. We start with the graph that consists of one node T, which is not processed and finish when all nodes of the graph are processed. This algorithm is a special case of the algorithm $\mathscr{A}_3$ considered in [1].

Algorithm $\mathscr{A}_{DAG}$ (construction of DAG $\Delta(T)$).

Input: A nonempty decision table T with n conditional attributes $f_1, \dots, f_n$.
Output: Directed acyclic graph $\Delta(T)$.

1. Construct the graph that consists of one node T, which is not marked as processed.
2. If all nodes of the graph are processed, then the algorithm halts and returns the resulting graph as $\Delta(T)$. Otherwise, choose a node (table) Θ that has not been processed yet.
3. If Θ is degenerate, then mark the node Θ as processed and proceed to step 2.
4. If Θ is not degenerate, then, for each $f_i \in E(\Theta)$, draw a bundle of edges from the node Θ. Let $E(\Theta, f_i) = \{a_1, \dots, a_k\}$. Then draw k edges from Θ and label these edges with systems of equations $\{f_i = a_1\}, \dots, \{f_i = a_k\}$. These edges enter nodes $\Theta\{f_i = a_1\}, \dots, \Theta\{f_i = a_k\}$, respectively. If some of the nodes $\Theta\{f_i = a_1\}, \dots, \Theta\{f_i = a_k\}$ are not present in the graph, then add these nodes to the graph. Mark the node Θ as processed and return to step 2.

The following statement about time complexity of the algorithm $\mathscr{A}_{DAG}$ follows immediately from Proposition 3.3 [1].

Proposition 3.1 *The time complexity of the algorithm $\mathscr{A}_{DAG}$ is bounded from above by a polynomial on the size of the input table T and the number $|SEP(T)|$ of different separable subtables of T.*

Generally, the time complexity of the algorithm $\mathscr{A}_{DAG}$ is exponential depending on the size of the input decision tables. Note that, in Sect. 3.4 of the book [1], classes of decision tables are described for each of which the number of separable subtables of decision tables from the class is bounded from above by a polynomial on the number of columns in the tables. For each of these classes, the time complexity of the algorithm $\mathscr{A}_{DAG}$ is polynomial depending on the size of the input decision tables.

3.2 Minimizing the Depth

Let T be a nonempty decision table with n conditional attributes $f_1, \ldots, f_n$. We can use the DAG $\Delta(T)$ to compute values $h^{(1)}(T), \ldots, h^{(5)}(T)$. Let $k \in \{1, \ldots, 5\}$. To find the value $h^{(k)}(T)$, for each node Θ of the DAG $\Delta(T)$, we compute the value $h^{(k)}(\Theta)$. It will be convenient for us to consider not only subtables that are nodes of $\Delta(T)$ but also empty subtable Λ of T and subtables T_r that contain only one row r of T and are not nodes of $\Delta(T)$. We begin with these special subtables and *terminal* nodes of $\Delta(T)$ (nodes without leaving edges) that are degenerate separable subtables of T and step-by-step move to the table T.

Let Θ be a terminal node of $\Delta(T)$ or $\Theta = T_r$ for some row r of T. Then $h^{(k)}(\Theta) = 0$: the decision tree that contains only one node labeled with the decision attached to all rows of Θ is a decision tree for Θ. If $\Theta = \Lambda$, then $h^{(k)}(\Theta) = 0$: the decision tree that contains only one node labeled with 0 will be considered as a decision tree for Λ.

Let Θ be a nonterminal node of $\Delta(T)$ such that, for each child Θ' of Θ, we already know the value $h^{(k)}(\Theta')$. Based on this information, we can find the minimum depth of a decision tree for Θ, which uses for the subtables corresponding to the children of the root decision trees of type k and in which the root is labeled

- With an attribute from $F(T)$ (we denote by $h_a^{(k)}(\Theta)$ the minimum depth of such a decision tree).
- With a hypothesis over T (we denote by $h_h^{(k)}(\Theta)$ the minimum depth of such a decision tree).
- With a proper hypothesis over T (we denote by $h_p^{(k)}(\Theta)$ the minimum depth of such a decision tree).

Since Θ is nondegenerate, the set $E(\Theta)$ is nonempty. We now describe three procedures for computing the values $h_a^{(k)}(\Theta)$, $h_h^{(k)}(\Theta)$, and $h_p^{(k)}(\Theta)$, respectively.

Let us consider a decision tree $\Gamma(f_i)$ for Θ in which the root is labeled with an attribute $f_i \in E(\Theta)$. For each $\delta \in E(T, f_i)$, there is an edge that leaves the root and enters a node $v(\delta)$. This edge is labeled with the equation system $\{f_i = \delta\}$. The node $v(\delta)$ is the root of a decision tree of type k for $\Theta\{f_i = \delta\}$ for which the depth is equal to $h^{(k)}(\Theta\{f_i = \delta\})$. It is clear that

$$h(\Gamma(f_i)) = 1 + \max\{h^{(k)}(\Theta\{f_i = \delta\}) : \delta \in E(T, f_i)\} .$$

Since $h^{(k)}(\Theta\{f_i = \delta\}) = h^{(k)}(\Lambda) = 0$ for any $\delta \in E(T, f_i) \setminus E(\Theta, f_i)$,

$$h(\Gamma(f_i)) = 1 + \max\{h^{(k)}(\Theta\{f_i = \delta\}) : \delta \in E(\Theta, f_i)\} . \tag{3.1}$$

Evidently, for any $\delta \in E(\Theta, f_i)$, the subtable $\Theta\{f_i = \delta\}$ is a child of Θ in the DAG $\Delta(T)$, i.e., we know the value $h^{(k)}(\Theta\{f_i = \delta\})$.

One can show that $h(\Gamma(f_i))$ is the minimum depth of a decision tree for Θ in which the root is labeled with the attribute f_i and which uses for the subtables corresponding to the children of the root decision trees of type k.

We should not consider attributes $f_i \in F(T) \setminus E(\Theta)$ since, for each such attribute, there is $\delta \in E(T, f_i)$ with $\Theta\{f_i = \delta\} = \Theta$, i.e., based on this attribute we cannot construct an optimal decision tree for Θ. As a result, we have

$$h_a^{(k)}(\Theta) = \min\{h(\Gamma(f_i)) : f_i \in E(\Theta)\} . \tag{3.2}$$

Computation of $h_a^{(k)}(\Theta)$. Construct the set of attributes $E(\Theta)$. For each attribute $f_i \in E(\Theta)$, compute the value $h(\Gamma(f_i))$ using (3.1). Compute the value $h_a^{(k)}(\Theta)$ using (3.2).

Remark 3.1 Let Θ be a nonterminal node of the DAG $\Delta(T)$ such that, for each child Θ' of Θ, we already know the value $h^{(k)}(\Theta')$. Then the procedure of computation of the value $h_a^{(k)}(\Theta)$ has polynomial time complexity depending on the size of decision table T.

A hypothesis $H = \{f_1 = \delta_1, \ldots, f_n = \delta_n\}$ over T is called *admissible for Θ and an attribute* $f_i \in F(T) = \{f_1, \ldots, f_n\}$ if, for any $\sigma \in E(T, f_i) \setminus \{\delta_i\}$, $\Theta\{f_i = \sigma\} \neq \Theta$. The hypothesis H is not admissible for Θ and an attribute $f_i \in F(T)$ if and only if $|E(\Theta, f_i)| = 1$ and $\delta_i \notin E(\Theta, f_i)$. The hypothesis H is called *admissible for Θ* if it is admissible for Θ and any attribute $f_i \in F(T)$.

Let us consider a decision tree $\Gamma(H)$ for Θ in which the root is labeled with an admissible for Θ hypothesis $H = \{f_1 = \delta_1, \ldots, f_n = \delta_n\}$. The set of answers for the query corresponding to the hypothesis H is equal to $A(H) = \{H, \{f_1 = \sigma_1\}, \ldots, \{f_n = \sigma_n\} : \sigma_1 \in E(T, f_1) \setminus \{\delta_1\}, \ldots, \sigma_n \in E(T, f_n) \setminus \{\delta_n\}\}$. For each $S \in A(H)$, there is an edge that leaves the root of $\Gamma(H)$ and enters a node $v(S)$. This edge is labeled with the equation system S. The node $v(S)$ is the root of a decision tree of type k for ΘS, which depth is equal to $h^{(k)}(\Theta S)$. It is clear that

$$h(\Gamma(H)) = 1 + \max\{h^{(k)}(\Theta S) : S \in A(H)\} .$$

We have $\Theta H = \Lambda$ or $\Theta H = T_r$ for some row r of T. Therefore $h^{(k)}(\Theta H) = 0$. Since H is admissible for Θ, $E(\Theta, f_i) \setminus \{\delta_i\} = \emptyset$ for any attribute $f \in F(T) \setminus E(\Theta)$. It is clear that $\Theta\{f_i = \sigma\} = \Lambda$ and $h^{(k)}(\Theta\{f_i = \sigma\}) = 0$ for any attribute $f_i \in E(\Theta)$ and any $\sigma \in E(T, f_i) \setminus \{\delta_i\}$ such that $\sigma \notin E(\Theta, f_i)$. Therefore

$$h(\Gamma(H)) = 1 + \max\{h^{(k)}(\Theta\{f_i = \sigma\}) : f_i \in E(\Theta), \sigma \in E(\Theta, f_i) \setminus \{\delta_i\}\} . \quad (3.3)$$

It is clear that, for any $f_i \in E(\Theta)$ and any $\sigma \in E(\Theta, f_i) \setminus \{\delta_i\}$, the subtable $\Theta\{f_i = \sigma\}$ is a child of Θ in the DAG $\Delta(T)$, i.e., we know the value $h^{(k)}(\Theta\{f_i = \sigma\})$.

One can show that $h(\Gamma(H))$ is the minimum depth of a decision tree for Θ in which the root is labeled with the hypothesis H and which uses for the subtables corresponding to the children of the root decision trees of type k.

We should not consider hypotheses that are not admissible for Θ since, for each such hypothesis H for corresponding query, there is an answer $S \in A(H)$ with $\Theta S = \Theta$, i.e., based on this hypothesis we cannot construct an optimal decision tree for Θ.

Computation of $h_h^{(k)}(\Theta)$. First, we construct a hypothesis

$$H_\Theta = \{f_1 = \delta_1, \ldots, f_n = \delta_n\}$$

for Θ. Let $f_i \in F(T) \setminus E(\Theta)$. Then δ_i is equal to the only number in the set $E(\Theta, f_i)$. Let $f_i \in E(\Theta)$. Then δ_i is the minimum number from $E(\Theta, f_i)$ for which $h^{(k)}(\Theta\{f_i = \delta_i\}) = \max\{h^{(k)}(\Theta\{f_i = \sigma\}) : \sigma \in E(\Theta, f_i)\}$. It is clear that H_Θ is admissible for Θ. Compute the value $h(\Gamma(H_\Theta))$ using (3.3). Simple analysis of (3.3) shows that $h(\Gamma(H_\Theta)) = h_h^{(k)}(\Theta)$.

Remark 3.2 Let Θ be a nonterminal node of the DAG $\Delta(T)$ such that, for each child Θ' of Θ, we already know the value $h^{(k)}(\Theta')$. Then the procedure of computation of the value $h_h^{(k)}(\Theta)$ has polynomial time complexity depending on the size of decision table T.

Computation of $h_p^{(k)}(\Theta)$. For each row $r = (\delta_1, \ldots, \delta_n)$ of the decision table T, we check if the corresponding proper hypothesis $H_r = \{f_1 = \delta_1, \ldots, f_n = \delta_n\}$ is admissible for Θ. For each admissible for Θ proper hypothesis $H_r = \{f_1 = \delta_1, \ldots, f_n = \delta_n\}$, we compute the value $h(\Gamma(H_r))$ using (3.3). One can show that the minimum among the obtained numbers is equal to $h_p^{(k)}(\Theta)$.

Remark 3.3 Let Θ be a nonterminal node of the DAG $\Delta(T)$ such that, for each child Θ' of Θ, we already know the value $h^{(k)}(\Theta')$. Then the procedure of computation of the value $h_p^{(k)}(\Theta)$ has polynomial time complexity depending on the size of decision table T.

We now describe an algorithm $\mathscr{A}_h$ that, for a given nonempty decision table T and given $k \in \{1, \ldots, 5\}$, calculates the value $h^{(k)}(T)$, which is the minimum depth of a decision tree of type k for the table T. During the work of this algorithm, we find for each node Θ of the DAG $\Delta(T)$ the value $h^{(k)}(\Theta)$.

Algorithm $\mathscr{A}_h$ (computation of $h^{(k)}(T)$).

Input: A nonempty decision table T, the directed acyclic graph $\Delta(T)$, and number $k \in \{1, \ldots, 5\}$.
Output: The value $h^{(k)}(T)$.

1. If a number is attached to each node of the DAG $\Delta(T)$, then return the number attached to the node T as $h^{(k)}(T)$ and halt the algorithm. Otherwise, choose a node Θ of the graph $\Delta(T)$ without attached number, which is either a terminal node of $\Delta(T)$ or a nonterminal node of $\Delta(T)$ for which all children have attached numbers.
2. If Θ is a terminal node, then attach to it the number $h^{(k)}(\Theta) = 0$ and proceed to step 1.
3. If Θ is not a terminal node, then depending on the value k do the following:

   - In the case $k = 1$, compute the value $h_a^{(1)}(\Theta)$ and attach to Θ the value $h^{(1)}(\Theta) = h_a^{(1)}(\Theta)$.
   - In the case $k = 2$, compute the value $h_h^{(2)}(\Theta)$ and attach to Θ the value $h^{(2)}(\Theta) = h_h^{(2)}(\Theta)$.
   - In the case $k = 3$, compute the values $h_a^{(3)}(\Theta)$ and $h_h^{(3)}(\Theta)$, and attach to Θ the value $h^{(3)}(\Theta) = \min\{h_a^{(3)}(\Theta), h_h^{(3)}(\Theta)\}$.
   - In the case $k = 4$, compute the value $h_p^{(4)}(\Theta)$ and attach to Θ the value $h^{(4)}(\Theta) = h_p^{(4)}(\Theta)$.
   - In the case $k = 5$, compute the values $h_a^{(5)}(\Theta)$ and $h_p^{(5)}(\Theta)$, and attach to Θ the value $h^{(5)}(\Theta) = \min\{h_a^{(5)}(\Theta), h_p^{(5)}(\Theta)\}$.

Proceed to step 1.

Using Remarks 3.1–3.3 one can prove the following statement.

Proposition 3.2 *The time complexity of the algorithm $\mathscr{A}_h$ is bounded from above by a polynomial on the size of the input table T and the number $|SEP(T)|$ of different separable subtables of T.*

3.3 Minimizing the Number of Realizable Nodes

Let T be a nonempty decision table with n conditional attributes $f_1, \ldots, f_n$. We can use the DAG $\Delta(T)$ to compute values $L^{(1)}(T), \ldots, L^{(5)}(T)$. Let $k \in \{1, \ldots, 5\}$. To find the value $L^{(k)}(T)$, we compute the value $L^{(k)}(\Theta)$ for each node Θ of the DAG $\Delta(T)$. We will consider not only subtables that are nodes of $\Delta(T)$ but also empty subtable Λ of T and subtables T_r that contain only one row r of T and are not nodes of $\Delta(T)$. We begin with these special subtables and terminal nodes of $\Delta(T)$ and step-by-step move to the table T.

Let Θ be a terminal node of $\Delta(T)$ or $\Theta = T_r$ for some row r of T. Then $L^{(k)}(\Theta) = 1$: the decision tree that contains only one node labeled with the decision attached to all rows of Θ is a decision tree for Θ. The only node of this tree is realizable relative to Θ. If $\Theta = \Lambda$, then $L^{(k)}(\Theta) = 0$: the decision tree that contains only one node labeled with 0 will be considered as a decision tree for Λ. The only node of this tree is not realizable relative to Λ.

Let Θ be a nonterminal node of $\Delta(T)$ such that, for each child Θ' of Θ, we already know the value $L^{(k)}(\Theta')$. Based on this information, we can find the minimum number of realizable relative to Θ nodes in a decision tree for Θ, which uses for the subtables corresponding to the children of the root decision trees of type k and in which the root is labeled

- With an attribute from $F(T)$ (we denote by $L_a^{(k)}(\Theta)$ the minimum number of realizable relative to Θ nodes in such a decision tree).
- With a hypothesis over T (we denote by $L_h^{(k)}(\Theta)$ the minimum number of realizable relative to Θ nodes in such a decision tree).
- With a proper hypothesis over T (we denote by $L_p^{(k)}(\Theta)$ the minimum number of realizable relative to Θ nodes in such a decision tree).

We now describe three procedures for computing the values $L_a^{(k)}(\Theta)$, $L_h^{(k)}(\Theta)$, and $L_p^{(k)}(\Theta)$, respectively. Since Θ is nondegenerate, the set $E(\Theta)$ is nonempty.

Let us consider a decision tree $\Gamma(f_i)$ for Θ in which the root is labeled with an attribute $f_i \in E(\Theta)$. For each $\delta \in E(T, f_i)$, there is an edge that leaves the root and enters a node $v(\delta)$. This edge is labeled with the equation system $\{f_i = \delta\}$. The node $v(\delta)$ is the root of a decision tree of type k for $\Theta\{f_i = \delta\}$ for which the number of realizable relative to $\Theta\{f_i = \delta\}$ nodes is equal to $L^{(k)}(\Theta\{f_i = \delta\})$. It is clear that $L(\Theta, \Gamma(f_i)) = 1 + \sum_{\delta \in E(T, f_i)} L^{(k)}(\Theta\{f_i = \delta\})$. Since $L^{(k)}(\Theta\{f_i = \delta\}) = L^{(k)}(\Lambda) = 0$ for any $\delta \in E(T, f_i) \setminus E(\Theta, f_i)$,

$$L(\Theta, \Gamma(f_i)) = 1 + \sum_{\delta \in E(\Theta, f_i)} L^{(k)}(\Theta\{f_i = \delta\}) . \qquad (3.4)$$

Evidently, for any $\delta \in E(\Theta, f_i)$, the subtable $\Theta\{f_i = \delta\}$ is a child of Θ in the DAG $\Delta(T)$, i.e., we know the value $L^{(k)}(\Theta\{f_i = \delta\})$. One can show that $L(\Theta, \Gamma(f_i))$ is the minimum number of realizable relative to Θ nodes in a decision tree for Θ, which uses for the subtables corresponding to the children of the root decision trees of type k and in which the root is labeled with the attribute f_i.

We should not consider attributes $f_i \in F(T) \setminus E(\Theta)$ since, for each such attribute, there is $\delta \in E(T, f_i)$ with $\Theta\{f_i = \delta\} = \Theta$, i.e., based on this attribute we cannot construct an optimal decision tree for Θ. As a result, we have

$$L_a^{(k)}(\Theta) = \min\{L(\Theta, \Gamma(f_i)) : f_i \in E(\Theta)\} . \tag{3.5}$$

Computation of $L_a^{(k)}(\Theta)$. Construct the set of attributes $E(\Theta)$. For each attribute $f_i \in E(\Theta)$, compute the value $L(\Theta, \Gamma(f_i))$ using (3.4). Compute the value $L_a^{(k)}(\Theta)$ using (3.5).

Let us consider a decision tree $\Gamma(H)$ for Θ in which the root is labeled with an admissible for Θ hypothesis $H = \{f_1 = \delta_1, \ldots, f_n = \delta_n\}$. For each $S \in A(H)$, there is an edge that leaves the root of $\Gamma(H)$ and enters a node $v(S)$. This edge is labeled with the equation system S. The node $v(S)$ is the root of a decision tree of type k for ΘS, for which the number of realizable relative to ΘS nodes is equal to $L^{(k)}(\Theta S)$. It is clear that $L(\Theta, \Gamma(H)) = 1 + \sum_{S \in A(H)} L^{(k)}(\Theta S)$.

Denote $r = (\delta_1, \ldots, \delta_n)$. It is easy to show that $\Theta H = \Lambda$ if r is not a row of Θ and $\Theta H = T_r$ if r is a row of Θ. Therefore

$$L^{(k)}(\Theta H) = \begin{cases} 0, \text{ if } r \text{ is not a row of } \Theta, \\ 1, \text{ if } r \text{ is a row of } \Theta. \end{cases} \tag{3.6}$$

Since H is admissible for Θ, $E(\Theta, f_i) \setminus \{\delta_i\} = \emptyset$ for any attribute $f_i \in F(T) \setminus E(\Theta)$. It is clear that $\Theta\{f_i = \sigma\} = \Lambda$ and $L^{(k)}(\Theta\{f_i = \sigma\}) = 0$ for any attribute $f_i \in E(\Theta)$ and any $\sigma \in E(T, f_i) \setminus \{\delta_i\}$ such that $\sigma \notin E(\Theta, f_i)$. Therefore

$$L(\Theta, \Gamma(H)) = L^{(k)}(\Theta H) + K(\Theta, H) , \tag{3.7}$$

where

$$K(\Theta, H) = 1 + \sum_{f_i \in E(\Theta), \sigma \in E(\Theta, f_i) \setminus \{\delta_i\}} L^{(k)}(\Theta\{f_i = \sigma\}) . \tag{3.8}$$

Evidently, for any $f_i \in E(\Theta)$ and any $\sigma \in E(\Theta, f_i) \setminus \{\delta_i\}$, the subtable $\Theta\{f_i = \sigma\}$ is a child of Θ in the DAG $\Delta(T)$, i.e., we know the value $L^{(k)}(\Theta\{f_i = \sigma\})$. It is easy to show that $L(\Theta, \Gamma(H))$ is the minimum number of realizable relative to Θ nodes in a decision tree for Θ, which uses for the subtables corresponding to the children of the root decision trees of type k and in which the root is labeled with the admissible for Θ hypothesis H.

We should not consider hypotheses that are not admissible for Θ since, for each such hypothesis H for corresponding query, there is an answer $S \in A(H)$ with $\Theta S = \Theta$, i.e., based on this hypothesis we cannot construct an optimal decision tree for Θ. As a result, we have

$$L_h^{(k)}(\Theta) = \min\{L(\Theta, \Gamma(H)) : H \in Adm(\Theta)\} , \tag{3.9}$$

where $Adm(\Theta)$ is the set of admissible hypotheses for Θ.

For each $f_i \in \{f_1, \ldots, f_n\}$, denote $a_i(\Theta) = \max\{L^{(k)}(\Theta\{f_i = \sigma\}) : \sigma \in E(\Theta, f_i)\}$ and $C(\Theta, f_i) = \{\sigma \in E(\Theta, f_i) : L^{(k)}(\Theta\{f_i = \sigma\}) = a_i(\Theta)\}$. Set $C(\Theta) = C(\Theta, f_1) \times \cdots \times C(\Theta, f_n)$. It is clear that, for each $\bar{\delta} = (\delta_1, \ldots, \delta_n) \in C(\Theta)$, the hypothesis $H_{\bar{\delta}} = \{f_1 = \delta_1, \ldots, f_n = \delta_n\}$ is admissible for Θ. Simple analysis of (3.8) shows that the set $\{H_{\bar{\delta}} : \bar{\delta} \in C(\Theta)\}$ coincides with the set of admissible for Θ hypotheses H that minimize the value $K(\Theta, H)$. Denote $K_{\min} = K(\Theta, H_{\bar{\delta}})$, where $\bar{\delta} \in C(\Theta)$.

Let there be a tuple $\bar{\delta} \in C(\Theta)$, which is not a row of Θ. Then $L^{(k)}(\Theta H_{\bar{\delta}}) = 0$ and $L_h^{(k)}(\Theta) = K_{\min}$. Let all tuples from $C(\Theta)$ be rows of Θ. We now show that $L_h^{(k)}(\Theta) = 1 + K_{\min}$. For any $\bar{\delta} \in C(\Theta)$, we have $L(\Theta, \Gamma(H_{\bar{\delta}})) = 1 + K_{\min}$. Therefore $L_h^{(k)}(\Theta) \leq 1 + K_{\min}$. Let us assume that $L_h^{(k)}(\Theta) < 1 + K_{\min}$. Then, by (3.9), there exists an admissible for Θ hypothesis $H = \{f_1 = \sigma_1, \ldots, f_n = \sigma_n\}$ for which $(\sigma_1, \ldots, \sigma_n) \notin C(\Theta)$ and $L(\Theta, \Gamma(H)) < 1 + K_{\min}$, but this is impossible since, according to (3.7), $L(\Theta, \Gamma(H)) \geq K(\Theta, H) \geq K_{\min} + 1$.

As a result, we have $L_h^{(k)}(\Theta) = K_{\min}$ if not all tuples from $C(\Theta)$ are rows of Θ, and $L_h^{(k)}(\Theta) = 1 + K_{\min}$ if all tuples from $C(\Theta)$ are rows of Θ.

Computation of $L_h^{(k)}(\Theta)$. For each $f_i \in \{f_1, \ldots, f_n\}$, we compute the value $a_i(\Theta) = \max\{L^{(k)}(\Theta\{f_i = \sigma\}) : \sigma \in E(\Theta, f_i)\}$ and construct set $C(\Theta, f_i) = \{\sigma \in E(\Theta, f_i) : L^{(k)}(\Theta\{f_i = \sigma\}) = a_i(\Theta)\}$. For a tuple $\bar{\delta} \in C(\Theta) = C(\Theta, f_1) \times \cdots \times C(\Theta, f_n)$, using (3.8) we compute the value $K_{\min} = K(\Theta, H_{\bar{\delta}})$. Then we count the number N of rows from Θ that belong to the set $C(\Theta)$ and compute the cardinality $|C(\Theta)|$ of the set $C(\Theta)$ that is equal to $|C(\Theta, f_1)| \cdot \ldots \cdot |C(\Theta, f_n)|$. As a result, we have $L_h^{(k)}(\Theta) = K_{\min}$ if $N < |C(\Theta)|$ and $L_h^{(k)}(\Theta) = 1 + K_{\min}$ if $N = |C(\Theta)|$.

Computation of $L_p^{(k)}(\Theta)$. For each row $r = (\delta_1, \ldots, \delta_n)$ of the decision table T, we check if the corresponding proper hypothesis $H_r = \{f_1 = \delta_1, \ldots, f_n = \delta_n\}$ is admissible for Θ. For each admissible for Θ proper hypothesis H_r, we compute the value $L(\Theta, \Gamma(H_r))$ using (3.6)–(3.8). One can show that the minimum among the obtained numbers is equal to $L_p^{(k)}(\Theta)$.

We now consider an algorithm $\mathscr{A}_L$ that, for a given nonempty decision table T and number $k \in \{1, \ldots, 5\}$, calculates the value $L^{(k)}(T)$, which is the minimum number of nodes realizable relative to T in a decision tree of type k for the table T. During the work of this algorithm, we find for each node Θ of the DAG $\Delta(T)$ the value $L^{(k)}(\Theta)$.

The description of the algorithm $\mathscr{A}_L$ is similar to the description of the algorithm $\mathscr{A}_h$. Instead of $h^{(k)}$ we should use $L^{(k)}$. For each $b \in \{a, h, p\}$, instead of $h_b^{(k)}$ we should use $L_b^{(k)}$. In particular, for each terminal node Θ, $L^{(k)}(\Theta) = 1$.

One can show that the procedures of computation of the values $L_a^{(k)}(\Theta)$, $L_h^{(k)}(\Theta)$, and $L_p^{(k)}(\Theta)$ have polynomial time complexity depending on the size of the decision table T. Using this fact, one can prove the following statement.

Proposition 3.3 *The time complexity of the algorithm $\mathscr{A}_L$ is bounded from above by a polynomial on the size of the input table T and the number $|SEP(T)|$ of different separable subtables of T.*

3.4 Minimizing the Number of Realizable Terminal Nodes

The procedure considered in this section is similar to the procedure of the minimization of the number of realizable nodes. The main difference is that in decision trees with the minimum number of realizable terminal nodes it is possible to meet constant attributes and hypotheses that are not admissible. Fortunately, for any decision table and any type of decision trees, there is a decision tree of this type with the minimum number of realizable terminal nodes for the considered table that do not use such attributes and hypotheses. We will omit many details and describe main steps only.

Let T be a nonempty decision table with n conditional attributes $f_1, \ldots, f_n$ and $k \in \{1, \ldots, 5\}$. To find the value $L_t^{(k)}(T)$, we compute the value $L_t^{(k)}(\Theta)$ for each node Θ of the DAG $\Delta(T)$. We begin with terminal nodes of $\Delta(T)$ that are degenerate separable subtables of T and step-by-step move to the table T.

Let Θ be a terminal node of $\Delta(T)$. Then $L_t^{(k)}(\Theta) = 1$: the decision tree that contains only one node labeled with the decision attached to all rows of Θ is a decision tree for Θ. The only node of this tree is a terminal node realizable relative to Θ.

Let Θ be a nonterminal node of $\Delta(T)$ such that, for each child Θ' of Θ, we already know the value $L_t^{(k)}(\Theta')$. Based on this information, we can find the minimum number of realizable relative to Θ terminal nodes in a decision tree for Θ, which uses for the subtables corresponding to children of the root decision trees of type k and in which the root is labeled

- With an attribute from $F(T)$ (we denote by $L_{t,a}^{(k)}(\Theta)$ the minimum number of realizable relative to Θ terminal nodes in such a decision tree).
- With a hypothesis over T (we denote by $L_{t,h}^{(k)}(\Theta)$ the minimum number of realizable relative to Θ terminal nodes in such a decision tree).
- With a proper hypothesis over T (we denote by $L_{t,p}^{(k)}(\Theta)$ the minimum number of realizable relative to Θ terminal nodes in such a decision tree).

We now describe three procedures for computing the values $L_{t,a}^{(k)}(\Theta)$, $L_{t,h}^{(k)}(\Theta)$, and $L_{t,p}^{(k)}(\Theta)$, respectively. Since Θ is nondegenerate, the set $E(\Theta)$ is nonempty.

Computation of $L_{t,a}^{(k)}(\Theta)$. Construct the set of attributes $E(\Theta)$. For each attribute $f_i \in E(\Theta)$, compute the value

$$L_{t,a}^{(k)}(\Theta, f_i) = \sum_{\delta \in E(\Theta, f_i)} L_t^{(k)}(\Theta\{f_i = \delta\}) .$$

Then compute the value

$$L_{t,a}^{(k)}(\Theta) = \min\{L_{t,a}^{(k)}(\Theta, f_i) : f_i \in E(\Theta)\} .$$

Computation of $L_{t,h}^{(k)}(\Theta)$. For each $f_i \in \{f_1, \ldots, f_n\}$, we compute the value $a_i(\Theta) = \max\{L_t^{(k)}(\Theta\{f_i = \sigma\}) : \sigma \in E(\Theta, f_i)\}$ and construct set $C(\Theta, f_i) = \{\sigma \in E(\Theta, f_i) : L_t^{(k)}(\Theta\{f_i = \sigma\}) = a_i(\Theta)\}$. For a tuple $(\delta_1, \ldots, \delta_n) \in C(\Theta) = C(\Theta, f_1) \times \cdots \times C(\Theta, f_n)$, we compute the value

$$K_{\min} = \sum_{f_i \in E(\Theta), \sigma \in E(\Theta, f_i) \setminus \{\delta_i\}} L_t^{(k)}(\Theta\{f_i = \sigma\}) .$$

Then we count the number N of rows from Θ that belong to the set $C(\Theta)$ and compute the cardinality $|C(\Theta)|$ of the set $C(\Theta)$ that is equal to $|C(\Theta, f_1)| \cdot \ldots \cdot |C(\Theta, f_n)|$. As a result, we have $L_{t,h}^{(k)}(\Theta) = K_{\min}$ if $N < |C(\Theta)|$ and $L_{t,h}^{(k)}(\Theta) = 1 + K_{\min}$ if $N = |C(\Theta)|$.

Computation of $L_{t,p}^{(k)}(\Theta)$. For each row $r = (\delta_1, \ldots, \delta_n)$ of the decision table T, we check if the corresponding proper hypothesis $H_r = \{f_1 = \delta_1, \ldots, f_n = \delta_n\}$ is admissible for Θ. For each admissible for Θ proper hypothesis H_r, we compute the value

$$L_{t,p}^{(k)}(\Theta, H_r) = 1 + \sum_{f_i \in E(\Theta), \sigma \in E(\Theta, f_i) \setminus \{\delta_i\}} L_t^{(k)}(\Theta\{f_i = \sigma\}) .$$

One can show that the minimum among the obtained numbers is equal to $L_{t,p}^{(k)}(\Theta)$.

We now consider an algorithm $\mathscr{A}_{L_t}$ that, for a given nonempty decision table T and number $k \in \{1, \ldots, 5\}$, calculates the value $L_t^{(k)}(T)$, which is the minimum number of terminal nodes realizable relative to T in a decision tree of type k for the table T. During the work of this algorithm, we find for each node Θ of the DAG $\Delta(T)$ the value $L_t^{(k)}(\Theta)$.

The description of the algorithm $\mathscr{A}_{L_t}$ is similar to the description of the algorithm $\mathscr{A}_h$. Instead of $h^{(k)}$ we should use $L_t^{(k)}$. For each $b \in \{a, h, p\}$, instead of $h_b^{(k)}$ we should use $L_{t,b}^{(k)}$. In particular, for each terminal node Θ, $L_t^{(k)}(\Theta) = 1$.

One can show that the procedures of computation of the values $L_{t,a}^{(k)}(\Theta)$, $L_{t,h}^{(k)}(\Theta)$, and $L_{t,p}^{(k)}(\Theta)$ have polynomial time complexity depending on the size of the decision table T. Using this fact, one can prove the following statement.

Proposition 3.4 *The time complexity of the algorithm $\mathscr{A}_{L_t}$ is bounded from above by a polynomial on the size of the input table T and the number $|SEP(T)|$ of different separable subtables of T.*

3.5 Minimizing the Number of Working Nodes

The procedure considered in this section is similar to the procedure of the minimization of the depth. We will omit many details and describe main steps only.

Let T be a nonempty decision table with n conditional attributes $f_1, \ldots, f_n$ and $k \in \{1, \ldots, 5\}$. To find the value $L_w^{(k)}(T)$, we compute the value $L_w^{(k)}(\Theta)$ for each node Θ of the DAG $\Delta(T)$. We begin with terminal nodes of $\Delta(T)$ that are degenerate separable subtables of T and step-by-step move to the table T.

Let Θ be a terminal node of $\Delta(T)$. Then $L_t^{(k)}(\Theta) = 0$: the decision tree that contains only one node labeled with the decision attached to all rows of Θ is a decision tree for Θ. This tree has no working nodes.

Let Θ be a nonterminal node of $\Delta(T)$ such that, for each child Θ' of Θ, we already know the value $L_w^{(k)}(\Theta')$. Based on this information, we can find the minimum number of working nodes in a decision tree for Θ, which uses for the subtables corresponding to children of the root decision trees of type k and in which the root is labeled

- With an attribute from $F(T)$ (we denote by $L_{w,a}^{(k)}(\Theta)$ the minimum number of working nodes in such a decision tree).
- With a hypothesis over T (we denote by $L_{w,h}^{(k)}(\Theta)$ the minimum number of working nodes in such a decision tree).
- With a proper hypothesis over T (we denote by $L_{w,p}^{(k)}(\Theta)$ the minimum number of working nodes in such a decision tree).

We now describe three procedures for computing the values $L_{w,a}^{(k)}(\Theta)$, $L_{w,h}^{(k)}(\Theta)$, and $L_{w,p}^{(k)}(\Theta)$, respectively. Since Θ is nondegenerate, the set $E(\Theta)$ is nonempty.

Computation of $L_{w,a}^{(k)}(\Theta)$. Construct the set of attributes $E(\Theta)$. For each attribute $f_i \in E(\Theta)$, compute the value

$$L_{w,a}^{(k)}(\Theta, f_i) = \sum_{\delta \in E(\Theta, f_i)} L_w^{(k)}(\Theta\{f_i = \delta\}) .$$

Then compute the value

$$L_{w,a}^{(k)}(\Theta) = \min\{L_{w,a}^{(k)}(\Theta, f_i) : f_i \in E(\Theta)\} .$$

Computation of $L_{w,h}^{(k)}(\Theta)$. First, we construct a hypothesis

$$H_\Theta = \{f_1 = \delta_1, \ldots, f_n = \delta_n\}$$

for Θ. Let $f_i \in F(T) \setminus E(\Theta)$. Then δ_i is equal to the only number in the set $E(\Theta, f_i)$. Let $f_i \in E(\Theta)$. Then δ_i is the minimum number from $E(\Theta, f_i)$ for which $L_w^{(k)}(\Theta\{f_i = \delta_i\}) = \max\{L_w^{(k)}(\Theta\{f_i = \sigma\}) : \sigma \in E(\Theta, f_i)\}$. Then

$$L^{(k)}_{w,h}(\Theta) = 1 + \sum_{f_i \in E(\Theta), \sigma \in E(\Theta, f_i) \setminus \{\delta_i\}} L^{(k)}_w(\Theta\{f_i = \sigma\}) .$$

Computation of $L^{(k)}_{w,p}(\Theta)$. For each row $r = (\delta_1, \ldots, \delta_n)$ of the decision table T, we check if the corresponding proper hypothesis $H_r = \{f_1 = \delta_1, \ldots, f_n = \delta_n\}$ is admissible for Θ. For each admissible for Θ proper hypothesis H_r, we compute the value

$$L^{(k)}_{w,p}(\Theta, H_r) = 1 + \sum_{f_i \in E(\Theta), \sigma \in E(\Theta, f_i) \setminus \{\delta_i\}} L^{(k)}_w(\Theta\{f_i = \sigma\}) .$$

One can show that the minimum among the obtained numbers is equal to $L^{(k)}_{w,p}(\Theta)$.

We now consider an algorithm $\mathcal{A}_{L_w}$ that, for a given nonempty decision table T and $k \in \{1, \ldots, 5\}$, calculates the value $L^{(k)}_w(T)$, which is the minimum number of working nodes in a decision tree of type k for the table T. During the work of this algorithm, we find for each node Θ of the DAG $\Delta(T)$ the value $L^{(k)}_w(\Theta)$.

The description of the algorithm $\mathcal{A}_{L_w}$ is similar to the description of the algorithm $\mathcal{A}_h$. Instead of $h^{(k)}$ we should use $L^{(k)}_w$. For each $b \in \{a, h, p\}$, instead of $h^{(k)}_b$ we should use $L^{(k)}_{w,b}$. In particular, for each terminal node Θ, $L^{(k)}_w(\Theta) = 0$.

One can show that the procedures of computation of the values $L^{(k)}_{w,a}(\Theta)$, $L^{(k)}_{w,h}(\Theta)$, and $L^{(k)}_{w,p}(\Theta)$ have polynomial time complexity depending on the size of the decision table T. Using this fact, one can prove the following statement.

Proposition 3.5 *The time complexity of the algorithm $\mathcal{A}_{L_w}$ is bounded from above by a polynomial on the size of the input table T and the number $|SEP(T)|$ of different separable subtables of T.*

3.6 On Number of Realizable Terminal Nodes

Based on the results of experiments, we formulated the following hypothesis: $L^{(1)}_t(T) = L^{(3)}_t(T) = L^{(5)}_t(T)$ for any decision table T. In this section, we prove it. First, we consider a simple lemma.

Lemma 3.1 *Let T be a decision table and T' be a subtable of the table T. Then $L^{(3)}_t(T') \le L^{(3)}_t(T)$.*

Proof It is easy to prove the considered inequality if T' is degenerate. Let T' be nondegenerate and Γ be a decision tree of type 3 for T with the minimum number of realizable relative to T terminal nodes. Then the root r of Γ is a working node. It is clear that the table $T'S(\Gamma, r)$ is nondegenerate. For each working node v of Γ such that the table $T'S(\Gamma, v)$ is degenerate and the table $T'S(\Gamma, v')$ is nondegenerate, where v' is the parent of v, we do the following. We remove all nodes and edges of the subtree of Γ with the root v with the exception of the node v. If $T'S(\Gamma, v) = \Lambda$, then we label the node v with the number 0. If the subtable $T'S(\Gamma, v)$ is nonempty, then we label the node v with the decision attached to each row of this subtable. We denote by Γ' the obtained decision tree. One can show that Γ' is a decision tree of type 3 for the table T' and $L_t^{(3)}(T', \Gamma') \le L_t^{(3)}(T, \Gamma)$. Therefore $L_t^{(3)}(T') \le L_t^{(3)}(T)$. □

Proposition 3.6 *For any decision table T, the following equalities hold:*

$$L_t^{(1)}(T) = L_t^{(3)}(T) = L_t^{(5)}(T) .$$

Proof It is clear that $L_t^{(3)}(T) \le L_t^{(5)}(T) \le L_t^{(1)}(T)$ for any decision table T. To prove the considered statement, it is enough to show that $L_t^{(1)}(T) \le L_t^{(3)}(T)$ for any decision table T. We will prove this inequality by induction on the number of attributes in the set $E(T)$.

We now show that $L_t^{(1)}(T) \le L_t^{(3)}(T)$ for any decision table T with $|E(T)| = 0$. If $|E(T)| = 0$, then either the table T is empty or the table T contains one row. Let T be empty. In this case, the decision tree that contains only one node labeled with 0 is considered as a decision tree for T. The only node of this tree is not realizable relative to T. Therefore $L_t^{(1)}(T) = L_t^{(3)}(T) = 0$. Let T contain one row. In this case, the decision tree that contains only one node labeled with the decision attached to the row of T is a decision tree for T. The only node of this tree is realizable relative to T. Therefore $L_t^{(1)}(T) = L_t^{(3)}(T) = 1$.

Let $n \ge 1$ and, for any decision table T with $|E(T)| \le n - 1$, the inequality $L_t^{(1)}(T) \le L_t^{(3)}(T)$ hold. Let T be a decision table with $|E(T)| = n$ and T have $m \ge n$ columns labeled with the attributes $f_1, \dots, f_m$. Let, for the definiteness, $E(T) = \{f_1, \dots, f_n\}$. If T is a degenerate table, then, as it is easy to show, $L_t^{(1)}(T) = L_t^{(3)}(T) = 1$. Let T be nondegenerate.

We denote by Γ a decision tree of type 3 for the table T for which $L_t(T, \Gamma) = L_t^{(3)}(T)$ and Γ has the minimum number of nodes among such decision trees. One can show that the root of Γ is either labeled with an attribute from $E(T)$ or with a hypothesis over T that is admissible for T. We now prove that the tree Γ can be transformed into a decision tree Γ^* of type 1 for the table T such that $L_t(T, \Gamma^*) \le L_t^{(3)}(T)$.

Let the root of Γ be labeled with an attribute $f_i \in E(T)$. Then, for each $\sigma \in E(T, f_i)$, the root of Γ has a child v_σ such that $TS(\Gamma, v_\sigma) = T\{f_i = \sigma\}$ and the root of Γ has no other children. Since $f_i \in E(T)$, $|E(T\{f_i = \sigma\}| \le n - 1$. Using the inductive hypothesis we obtain that there is a decision tree Γ_σ of type 1 for the table $T\{f_i = \sigma\}$ such that $L_t(T\{f_i = \sigma\}, \Gamma_\sigma) \le L_t^{(3)}(T\{f_i = \sigma\})$. For each child v_σ of the root of Γ, we replace the subtree of Γ with the root v_σ with the tree Γ_σ. As a result, we obtain a decision tree Γ^* of type 1 for the table T such that $L_t(T, \Gamma^*) \le L_t(T, \Gamma) = L_t^{(3)}(T)$.

Let the root of Γ be labeled with a hypothesis $H = \{f_1 = \delta_1, \dots, f_m = \delta_m\}$ over T that is admissible for T—see Fig. 3.1, which depicts a prefix of the tree Γ. The root of Γ has a child v_0 such that $TS(\Gamma, v_0) = TH = T\{f_1 = \delta_1, \dots, f_n = \delta_n\}$. For each $f_i \in E(T)$ and each

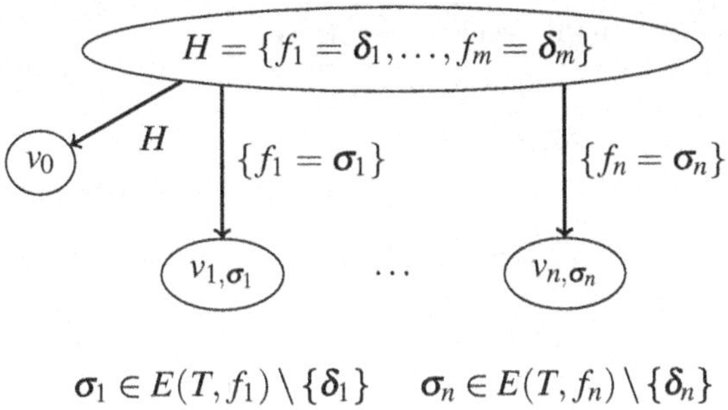

Fig. 3.1 Prefix of decision tree Γ

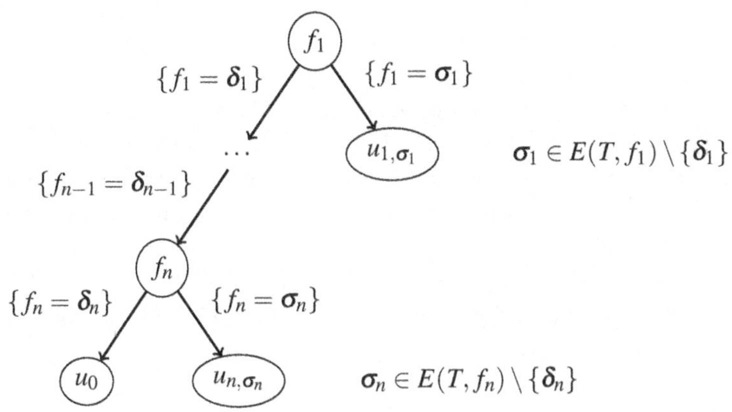

Fig. 3.2 Prefix of decision tree Γ^*

$\sigma_i \in E(T, f_i) \setminus \{\delta_i\}$, the root of Γ has a child v_{i,σ_i} such that $TS(\Gamma, v_{i,\sigma_i}) = T\{f_i = \sigma_i\}$. The root of Γ has no other children.

We transform the tree Γ into a decision tree Γ^* of type 1 for the table T—see Fig. 3.2, which depicts a prefix of the tree Γ^*. For the node u_0 of the considered prefix, $TS(\Gamma^*, u_0) = T\{f_1 = \delta_1, \ldots, f_n = \delta_n\} = TH$. For each $f_i \in E(T)$ and each $\sigma_i \in E(T, f_i) \setminus \{\delta_i\}$, the node of this prefix labeled with the attribute f_i has a child u_{i,σ_i} such that $TS(\Gamma^*, u_{i,\sigma_i}) = T\{f_1 = \delta_1, \ldots, f_{i-1} = \delta_{i-1}, f_i = \sigma_i\}$. It is clear that $TS(\Gamma^*, u_{i,\sigma_i})$ is a subtable of $TS(\Gamma, v_{i,\sigma_i})$. By Lemma 3.1, $L_t^{(3)}(TS(\Gamma^*, u_{i,\sigma_i})) \leq L_t^{(3)}(TS(\Gamma, v_{i,\sigma_i}))$. It is also clear that $|E(TS(\Gamma^*, u_{i,\sigma_i})| \leq n - 1$. Using the inductive hypothesis we obtain that there is a decision tree Γ_{i,σ_i} of type 1 for the table $TS(\Gamma^*, u_{i,\sigma_i})$ such that

$$L_t(TS(\Gamma^*, u_{i,\sigma_i}), \Gamma_{i,\sigma_i}) \leq L_t^{(3)}(TS(\Gamma^*, u_{i,\sigma_i})) \leq L_t^{(3)}(TS(\Gamma, v_{i,\sigma_i})) .$$

Table 3.1 Decision tables from [6] used in experiments

Decision table	Number of rows	Number of attributes
BALANCE-SCALE	625	5
BREAST-CANCER	266	10
CARS	1728	7
HAYES-ROTH-DATA	69	5
NURSERY	12,960	9
SOYBEAN-SMALL	47	36
TIC-TAC-TOE	958	10
ZOO-DATA	59	17

We now transform the prefix of a decision tree Γ^* depicted in Fig. 3.2 into a decision tree Γ^* of type 1 for the table T. First, we transform the node u_0 into a terminal node labeled with the number 0 if $(\delta_1, \ldots, \delta_n)$ is not a row of T and labeled with the decision attached to $(\delta_1, \ldots, \delta_n)$ if this tuple is a row of T. Next, for each $f_i \in E(T)$ and each $\sigma_i \in E(T, f_i) \setminus \{\delta_i\}$, we replace the node u_{i,σ_i} with the tree Γ_{i,σ_i}. It is clear that the obtained tree Γ^* is a decision tree of type 1 for the decision table T and $L_t(T, \Gamma^*) \le L_t(T, \Gamma) = L_t^{(3)}(T)$.

We proved that, for any decision table T, $L_t^{(1)}(T) \le L_t^{(3)}(T)$ and, hence, $L_t^{(1)}(T) = L_t^{(3)}(T) = L_t^{(5)}(T)$. □

3.7 Results of Experiments

We make experiments on eight decision tables from the UCI ML Repository [6]. Table 3.1 contains information about each of these decision tables: its name, the number of rows, and the number of attributes. For each of the considered four cost functions, each of the consider five types of decision trees, and each of the considered eight decision tables, we find the minimum cost of a decision tree of the given type for the given table.

For $n = 3, \ldots, 6$, we randomly generate 100 Boolean functions with n variables. We represent each Boolean function f with n variables $x_1, \ldots, x_n$ as a decision table T_f with n columns labeled with variables $x_1, \ldots, x_n$ considered as attributes and with 2^n rows that are all possible n-tuples of values of the variables. Each row is labeled with the decision that is the value of the function f on the corresponding n-tuple. We consider decision trees for the table T_f as decision trees computing the function f.

For each of the considered four cost functions, each of the considered five types of decision trees, and each of the generated Boolean functions, using its decision table representation, we find the minimum cost of a decision tree of the given type computing this function.

The following remarks clarify some experimental results considered later.

Table 3.2 Experimental results for decision tables from [6]—depth

Decision table, T	$h^{(1)}(T)$	$h^{(2)}(T)$	$h^{(3)}(T)$	$h^{(4)}(T)$	$h^{(5)}(T)$
BALANCE-SCALE	**4**	**4**	**4**	**4**	**4**
BREAST-CANCER	6	6	**5**	6	**5**
CARS	**6**	**6**	**6**	**6**	**6**
HAYES-ROTH-DATA	**4**	**4**	**4**	**4**	**4**
NURSERY	8	8	**7**	8	**7**
SOYBEAN-SMALL	**2**	4	**2**	6	**2**
TIC-TAC-TOE	6	6	**5**	8	6
ZOO-DATA	**4**	**4**	**4**	5	**4**
Average	5.00	5.25	4.63	5.88	4.75

From Proposition 3.6 it follows that $L_t^{(1)}(T) = L_t^{(3)}(T) = L_t^{(5)}(T)$ for any decision table T.

Let f be a Boolean function with $n \geq 1$ variables. Since each hypothesis over the decision table T_f is proper, the following equalities hold:

$$\begin{aligned} h^{(2)}(T_f) = h^{(4)}(T_f),\ \ h^{(3)}(T_f) = h^{(5)}(T_f)\ , \\ L^{(2)}(T_f) = L^{(4)}(T_f),\ \ L^{(3)}(T_f) = L^{(5)}(T_f)\ , \\ L_t^{(2)}(T_f) = L_t^{(4)}(T_f),\ \ L_t^{(3)}(T_f) = L_t^{(5)}(T_f)\ , \\ L_w^{(2)}(T_f) = L_w^{(4)}(T_f),\ \ L_w^{(3)}(T_f) = L_w^{(5)}(T_f)\ . \end{aligned}$$

3.7.1 Depth

Results of experiments on eight decision tables from [6] and the depth are represented in Table 3.2. The first column contains the name of the considered decision table T. The last five columns contain values $h^{(1)}(T), \ldots, h^{(5)}(T)$ (minimum values for each decision table are in bold).

Decision trees with the minimum depth using attributes (type 1) are optimal for 5 decision tables, using hypotheses (type 2) are optimal for 4 tables, using attributes and hypotheses (type 3) are optimal for 8 tables, using proper hypotheses (type 4) are optimal for 3 tables, using attributes and proper hypotheses (type 5) are optimal for 7 tables.

For the decision table SOYBEAN-SMALL, we must use attributes to construct an optimal decision tree. For this table, it is enough to use only attributes. For the decision tables BREAST-CANCER and NURSERY, we must use both attributes and hypotheses to construct optimal decision trees. For these tables, it is enough to use attributes and proper hypotheses.

Table 3.3 Experimental results for Boolean functions—depth

Number of variables, n	$h^{(1)}$	$h^{(2)}$	$h^{(3)}$	$h^{(4)}$	$h^{(5)}$
3	${}_{2}2.82_{3}$	${}_{1}2.06_{3}$	${}_{1}\mathbf{1.89}_{2}$	${}_{1}2.06_{3}$	${}_{1}\mathbf{1.89}_{2}$
4	${}_{3}3.94_{4}$	${}_{2}3.05_{4}$	${}_{2}\mathbf{2.97}_{3}$	${}_{2}3.05_{4}$	${}_{2}\mathbf{2.97}_{3}$
5	${}_{4}4.95_{5}$	${}_{4}4.08_{5}$	${}_{3}\mathbf{3.99}_{4}$	${}_{4}4.08_{5}$	${}_{3}\mathbf{3.99}_{4}$
6	${}_{5}5.99_{6}$	${}_{5}5.01_{6}$	${}_{5}\mathbf{5.00}_{5}$	${}_{5}5.01_{6}$	${}_{5}\mathbf{5.00}_{5}$

For the decision table TIC-TAC-TOE, we must use both attributes and hypotheses to construct optimal decision trees. For this table, it is not enough to use attributes and proper hypotheses.

Results of experiments on Boolean functions and the depth are represented in Table 3.3. The first column contains the number of variables in the considered Boolean functions. The last five columns contain information about values $h^{(1)}, \ldots, h^{(5)}$ in the format ${}_{\min}Avg_{\max}$ (minimum values of Avg for each n are in bold).

From the obtained results it follows that, generally, the decision trees of types 2 and 4 are better than the decision trees of type 1, and the decision trees of types 3 and 5 are better than the decision trees of types 2 and 4.

3.7.2 Number of Realizable Nodes

Results of experiments on eight decision tables from [6] and the number of realizable nodes are represented in Table 3.4. The first column contains the name of the considered decision table T. The last five columns contain values $L^{(1)}(T), \ldots, L^{(5)}(T)$ (minimum values for each decision table are in bold).

Decision trees with the minimum number of realizable nodes using attributes (type 1) are optimal for 4 decision tables, using hypotheses (type 2) are optimal for 0 tables, using attributes and hypotheses (type 3) are optimal for 8 tables, using proper hypotheses (type 4) are optimal for 0 tables, using attributes and proper hypotheses (type 5) are optimal for 8 tables.

Decision trees of types 3 and 5 can be a bit better than the decision trees of type 1. Decision trees of types 2 and 4 are far from the optimal.

For the decision tables HAYES-ROTH-DATA, SOYBEAN-SMALL, TIC-TAC-TOE, and ZOO-DATA, we must use attributes to construct optimal decision trees. For these tables, it is enough to use only attributes. For the rest of the considered decision tables, we must use both attributes and hypotheses to construct optimal decision trees. For these tables, it is enough to use attributes and proper hypotheses.

Results of experiments on Boolean functions and the number of realizable nodes are represented in Table 3.5. The first column contains the number of variables in the considered

Table 3.4 Experimental results for decision tables from [6]—number of realizable nodes

Decision table, T	$L^{(1)}(T)$	$L^{(2)}(T)$	$L^{(3)}(T)$	$L^{(4)}(T)$	$L^{(5)}(T)$
BALANCE-SCALE	501	5234	**499**	5234	**499**
BREAST-CANCER	161	18,061	**159**	24,226	**159**
CARS	396	65,250	**391**	65,250	**391**
HAYES-ROTH-DATA	**52**	317	**52**	338	**52**
NURSERY	1066	12,625,955	**1061**	12,625,955	**1061**
SOYBEAN-SMALL	**6**	4839	**6**	21,251	**6**
TIC-TAC-TOE	**244**	154,311	**244**	468,447	**244**
ZOO-DATA	**17**	1370	**17**	2847	**17**
Average	305	1,609,417	304	1,651,694	304

Table 3.5 Experimental results for Boolean functions—number of realizable nodes

Number of variables, n	$L^{(1)}$	$L^{(2)}$	$L^{(3)}$	$L^{(4)}$	$L^{(5)}$
3	$_{5}8.41_{13}$	$_{5}12.38_{22}$	$_{5}\mathbf{7.40}_{12}$	$_{5}12.38_{22}$	$_{5}\mathbf{7.40}_{12}$
4	$_{9}16.26_{25}$	$_{14}43.89_{66}$	$_{8}\mathbf{14.59}_{25}$	$_{14}43.89_{66}$	$_{8}\mathbf{14.59}_{25}$
5	$_{17}30.42_{41}$	$_{113}201.95_{283}$	$_{17}\mathbf{27.83}_{39}$	$_{113}201.95_{283}$	$_{17}\mathbf{27.83}_{39}$
6	$_{49}58.94_{77}$	$_{638}1,057.16_{1406}$	$_{46}\mathbf{54.13}_{71}$	$_{638}1,057.16_{1406}$	$_{46}\mathbf{54.13}_{71}$

Boolean functions. The last five columns contain information about values $L^{(1)}, \ldots, L^{(5)}$ in the format $_{\min}Avg_{\max}$ (minimum values of *Avg* for each n are in bold).

From the obtained results it follows that, generally, the decision trees of types 3 and 5 are slightly better than the decision trees of type 1, and the decision trees of types 2 and 4 are far from the optimal.

3.7.3 Number of Realizable Terminal Nodes

Results of experiments on eight decision tables from [6] and the number of realizable terminal nodes are represented in Table 3.6. The first column contains the name of the considered decision table T. The last five columns contain values $L_t^{(1)}(T), \ldots, L_t^{(5)}(T)$ (minimum values for each decision table are in bold).

Decision trees of types 1, 3, and 5 are optimal for each of the considered tables. Decision trees of types 2 and 4 are far from the optimal.

Results of experiments on Boolean functions and the number of realizable terminal nodes are represented in Table 3.7. The first column contains the number of variables in

Table 3.6 Experimental results for decision tables from [6]—number of realizable terminal nodes

Decision table, T	$L_t^{(1)}(T)$	$L_t^{(2)}(T)$	$L_t^{(3)}(T)$	$L_t^{(4)}(T)$	$L_t^{(5)}(T)$
BALANCE-SCALE	**401**	4369	**401**	4369	**401**
BREAST-CANCER	**99**	15,577	**99**	21,208	**99**
CARS	**290**	50,260	**290**	50,260	**290**
HAYES-ROTH-DATA	**36**	240	**36**	259	**36**
NURSERY	**759**	9,643,103	**759**	9,643,103	**759**
SOYBEAN-SMALL	**4**	4508	**4**	19,963	**4**
TIC-TAC-TOE	**155**	125,604	**155**	401,862	**155**
ZOO-DATA	**10**	1217	**10**	2560	**10**
Average	219	1,230,610	219	1,267,948	219

Table 3.7 Experimental results for Boolean functions—number of realizable terminal nodes

Number of variables, n	$L_t^{(1)}$	$L_t^{(2)}$	$L_t^{(3)}$	$L_t^{(4)}$	$L_t^{(5)}$
3	${}_{3}\mathbf{4.70}_{7}$	${}_{4}8.83_{14}$	${}_{3}\mathbf{4.70}_{7}$	${}_{4}8.83_{14}$	${}_{3}\mathbf{4.70}_{7}$
4	${}_{5}\mathbf{8.63}_{13}$	${}_{11}31.53_{45}$	${}_{5}\mathbf{8.63}_{13}$	${}_{11}31.53_{45}$	${}_{5}\mathbf{8.63}_{13}$
5	${}_{9}\mathbf{15.71}_{21}$	${}_{86}145.68_{200}$	${}_{9}\mathbf{15.71}_{21}$	${}_{86}145.68_{200}$	${}_{9}\mathbf{15.71}_{21}$
6	${}_{25}\mathbf{29.97}_{39}$	${}_{485}770.52_{1005}$	${}_{25}\mathbf{29.97}_{39}$	${}_{485}770.52_{1005}$	${}_{25}\mathbf{29.97}_{39}$

the considered Boolean functions. The last five columns contain information about values $L_t^{(1)}, \ldots, L_t^{(5)}$ in the format ${}_{\min}Avg_{\max}$ (minimum values of *Avg* for each n are in bold).

Decision trees of types 1, 3, and 5 are optimal for each of the considered Boolean functions, and decision trees of types 2 and 4 are, generally, far from the optimal.

3.7.4 Number of Working Nodes

Results of experiments on eight decision tables from [6] and the number of working nodes are represented in Table 3.8. The first column contains the name of the considered decision table T. The last five columns contain values $L_w^{(1)}(T), \ldots, L_w^{(5)}(T)$ (minimum values for each decision table are in bold).

Decision trees with the minimum number of working nodes using attributes (type 1) are optimal for 2 decision tables, using hypotheses (type 2) are optimal for 0 tables, using attributes and hypotheses (type 3) are optimal for 8 tables, using proper hypotheses (type 4) are optimal for 0 tables, using attributes and proper hypotheses (type 5) are optimal for 7 tables.

Table 3.8 Experimental results for decision tables from [6]—number of working nodes

Decision table, T	$L_w^{(1)}(T)$	$L_w^{(2)}(T)$	$L_w^{(3)}(T)$	$L_w^{(4)}(T)$	$L_w^{(5)}(T)$
BALANCE-SCALE	100	865	**98**	865	**98**
BREAST-CANCER	49	2415	**45**	2926	**45**
CARS	104	14,981	**99**	14,981	**99**
HAYES-ROTH-DATA	16	77	**14**	77	**14**
NURSERY	286	2,980,719	**281**	2,980,719	**281**
SOYBEAN-SMALL	**2**	330	**2**	1281	**2**
TIC-TAC-TOE	88	27,867	**85**	65,104	86
ZOO-DATA	**7**	151	**7**	284	**7**
Average	82	378,426	79	383,280	79

Table 3.9 Experimental results for Boolean functions—number of working nodes

Number of variables, n	$L_w^{(1)}$	$L_w^{(2)}$	$L_w^{(3)}$	$L_w^{(4)}$	$L_w^{(5)}$
3	${}_{2}3.70_{6}$	${}_{1}3.55_{8}$	${}_{1}\mathbf{2.58}_{4}$	${}_{1}3.55_{8}$	${}_{1}\mathbf{2.58}_{4}$
4	${}_{4}7.63_{12}$	${}_{3}12.36_{21}$	${}_{3}\mathbf{5.62}_{9}$	${}_{3}12.36_{21}$	${}_{3}\mathbf{5.62}_{9}$
5	${}_{8}14.71_{20}$	${}_{27}56.25_{83}$	${}_{8}\mathbf{11.38}_{15}$	${}_{27}56.25_{83}$	${}_{8}\mathbf{11.38}_{15}$
6	${}_{24}28.97_{38}$	${}_{153}286.52_{401}$	${}_{19}\mathbf{22.76}_{29}$	${}_{153}286.52_{401}$	${}_{19}\mathbf{22.76}_{29}$

Decision trees of types 3 and 5 can be a bit better than the decision trees of type 1. Decision trees of types 2 and 4 are far from the optimal.

For all decision tables with the exception of SOYBEAN-SMALL and ZOO-DATA, we must use both attributes and hypotheses to construct optimal decision trees. Moreover, for TIC-TAC-TOE it is not enough to use attributes and proper hypotheses. For SOYBEAN-SMALL and ZOO-DATA, it is enough to use only attributes to construct optimal decision trees.

Results of experiments on Boolean functions and the number of working nodes are represented in Table 3.9. The first column contains the number of variables in the considered Boolean functions. The last five columns contain information about values $L_w^{(1)}, \ldots, L_w^{(5)}$ in the format ${}_{\min}Avg_{\max}$ (minimum values of Avg for each n are in bold).

From the obtained results it follows that, generally, the decision trees of types 3 and 5 are better than the decision trees of type 1, and the decision trees of types 2 and 4 are far from the optimal.

We can now sum up the results of the experiments. Generally, the decision trees of types 3 and 5 are slightly better than the decision trees of type 1. Decision trees of types 2 and 4 have, generally, too many nodes.

3.8 Conclusions

In this chapter, we studied decision trees that can use queries based on one attribute each and queries based on hypotheses about values of all attributes. We designed dynamic programming algorithms for the minimization of four cost functions for such decision trees and considered results of computer experiments.

References

1. AbouEisha, H., Amin, T., Chikalov, I., Hussain, S., Moshkov, M.: Extensions of Dynamic Programming for Combinatorial Optimization and Data Mining, Intelligent Systems Reference Library, vol. 146. Springer (2019)
2. Alsolami, F., Azad, M., Chikalov, I., Moshkov, M.: Decision and Inhibitory Trees and Rules for Decision Tables with Many-valued Decisions, Intelligent Systems Reference Library, vol. 156. Springer (2020)
3. Azad, M., Chikalov, I., Hussain, S., Moshkov, M.: Minimizing depth of decision trees with hypotheses. In: Ramanna, S., Cornelis, C., Ciucci, D. (eds.) Rough Sets—International Joint Conference, IJCRS 2021, Bratislava, Slovakia, 19–24 Sept 2021, Lecture Notes in Computer Science, vol. 12872, pp. 123–133. Springer (2021)
4. Azad, M., Chikalov, I., Hussain, S., Moshkov, M.: Minimizing number of nodes in decision trees with hypotheses. In: Watrobski, J., Salabun, W., Toro, C., Zanni-Merk, C., Howlett, R.J., Jain, L.C. (eds.) 25th International Conference on Knowledge-Based and Intelligent Information & Engineering Systems, KES 2021, 8–10 Sept 2021, Szczecin, Poland, Procedia Computer Science, vol. 192, pp. 232–240. Elsevier (2021). https://doi.org/10.1016/j.procs.2021.08.024
5. Azad, M., Chikalov, I., Hussain, S., Moshkov, M.: Optimization of decision trees with hypotheses for knowledge representation. Electronics **10**(13), 1580 (2021). https://doi.org/10.3390/electronics10131580
6. Dua, D., Graff, C.: UCI Machine Learning Repository. University of California, Irvine, School of Information and Computer Sciences (2017). http://archive.ics.uci.edu/ml

Construction of Optimal Decision Trees and Deriving Decision Rules from Them

4

In Chap. 3, we studied five types of decision trees based on (1) attributes, (2) hypotheses, (3) attributes and hypotheses, (4) proper hypotheses, and (5) attributes and proper hypotheses. We considered four cost functions: the depth, the number of realizable nodes, the number of realizable terminal nodes, and the number of working (nonterminal) nodes. For each cost function, we proposed a dynamic programming algorithm that, for a given decision table and given type of decision trees, finds the minimum cost of a decision tree of the considered type for this table. The results of computer experiments showed that optimal decision trees of types 3 and 5 can have less complexity than the decision trees of type 1. Therefore they can be more suitable as tools for knowledge representation.

The present chapter, which is based on the paper [2], has two aims. The first aim is to construct optimal decision trees with hypotheses. The algorithms from Chap. 3 allow us only to find the complexity of optimal trees but do not allow to build such trees. The second aim is to study the length and coverage of decision rules derived from optimal decision trees. Decision rules can be considered as one of the simplest and most understandable models for knowledge representation. Deriving decision rules from decision trees is a well-known approach. We want to compare the decision rules derived from optimal decision trees with hypotheses with ones derived from optimal conventional decision trees.

In this chapter, we consider two cost functions: the depth and the number of working nodes. For each of these two cost functions, we propose a dynamic programming algorithm (based on the corresponding algorithm from Chap. 3) that, for a given decision table and given type of decision trees, constructs for this table a decision tree of the considered type with the minimum cost. We can use these algorithms to construct optimal decision trees for medium-sized decision tables.

Unfortunately, we cannot use similar approach to build optimal decision trees of types 2 and 3 relative to the number of realizable nodes and optimal decision trees of type 2 relative to the number of realizable terminal nodes as the optimization algorithms become too ineffective. In this chapter, we propose another way to construct optimal decision trees

M. Azad et al., *Decision Trees with Hypotheses*, Synthesis Lectures on Intelligent Technologies, https://doi.org/10.1007/978-3-031-08585-7_4

for the mentioned cases. However, we do not implement corresponding algorithms and do not make experiments on optimal trees relative to the considered two cost functions.

We construct decision trees of five types optimal relative to the depth and relative to the number of working nodes for eight decision tables from the UCI ML Repository [3] and for 100 randomly generated Boolean functions with n variables, where $n = 3, \ldots, 6$. We compare the length and coverage of decision rules derived from the constructed decision trees. For a decision tree with hypotheses for some rows of the considered decision table, it can be more than one derived decision rule that covers the row. In this case, for each row, we choose the best rule. The results of computer experiments show that the decision rules derived from decision trees with hypotheses in many cases are better than the ones derived from conventional decision trees.

The rest of the chapter is organized as follows. In Sects. 4.1 and 4.2, we consider dynamic programming algorithms for the construction of optimal decision trees relative to the depth and to the number of working nodes. In Sect. 4.3, we discuss how it is possible to construct optimal decision trees relative to the two remaining cost functions. Section 4.4 contains results of computer experiments and Sect. 4.5—short conclusions.

4.1 Construction of Decision Trees with Minimum Depth

Let T be a nonempty decision table with n conditional attributes $f_1, \ldots, f_n$ and $k \in \{1, \ldots, 5\}$. We can use the DAG $\Delta(T)$ (builded by the algorithm $\mathscr{A}_{DAG}$ described in Sect. 3.1) to construct a decision tree $\Gamma^{(k)}(T)$ of type k with the minimum depth for the decision table T. To construct the tree $\Gamma^{(k)}(T)$, for each node Θ of the DAG $\Delta(T)$, we construct a decision tree $\Gamma^{(k)}(\Theta)$ of type k with the minimum depth for the decision table Θ. It will be convenient for us to consider not only subtables that are nodes of $\Delta(T)$ but also empty subtable Λ of T and subtables T_r that contain only one row r of T and are not nodes of $\Delta(T)$. We begin with these special subtables and terminal nodes of $\Delta(T)$ (nodes without leaving edges) that are degenerate separable subtables of T and step-by-step move to the table T.

Let Θ be a terminal node of $\Delta(T)$ or $\Theta = T_r$ for some row r of T. Then $\Gamma^{(k)}(\Theta)$ is the decision tree that contains only one node labeled with the decision attached to all rows of Θ. If $\Theta = \Lambda$, then $\Gamma^{(k)}(\Theta)$ is the decision tree that contains only one node labeled with 0 (this tree will be considered as a decision tree for Λ).

Let Θ be a nonterminal node of $\Delta(T)$ such that, for each child Θ' of Θ, we already constructed a decision tree $\Gamma^{(k)}(\Theta')$. Based on these trees, we can construct a decision tree with the minimum depth for Θ, which uses for the subtables corresponding to the children of the root decision trees of type k and in which the root is labeled

- With an attribute from $F(T)$ (we denote by $\Gamma_a^{(k)}(\Theta)$ such a decision tree).
- With a hypothesis over T (we denote by $\Gamma_h^{(k)}(\Theta)$ such a decision tree).

- With a proper hypothesis over T (we denote by $\Gamma_p^{(k)}(\Theta)$ such a decision tree).

Since Θ is nondegenerate, the set $E(\Theta)$ is nonempty. We now describe three procedures for construction of the trees $\Gamma_a^{(k)}(\Theta)$, $\Gamma_h^{(k)}(\Theta)$, and $\Gamma_p^{(k)}(\Theta)$, respectively.

Let us consider a decision tree $\Gamma(f_i)$ for Θ in which the root is labeled with an attribute $f_i \in E(\Theta)$. For each $\delta \in E(T, f_i)$, there is an edge that leaves the root and enters a node $v(\delta)$. This edge is labeled with the equation system $\{f_i = \delta\}$. The node $v(\delta)$ is the root of the decision tree $\Gamma^{(k)}(\Theta\{f_i = \delta\})$. It is clear that

$$h(\Gamma(f_i)) = 1 + \max\{h(\Gamma^{(k)}(\Theta\{f_i = \delta\})) : \delta \in E(T, f_i)\} . \tag{4.1}$$

Using (4.1), one can show that $\Gamma(f_i)$ is a decision tree with the minimum depth for Θ in which the root is labeled with the attribute f_i and which uses for the subtables corresponding to the children of the root decision trees of type k.

We should not consider attributes $f_i \in F(T) \setminus E(\Theta)$ since, for each such attribute, there is $\delta \in E(T, f_i)$ with $\Theta\{f_i = \delta\} = \Theta$, i.e., based on this attribute we cannot construct an optimal decision tree for Θ.

Construction of the tree $\Gamma_a^{(k)}(\Theta)$. Construct the set of attributes $E(\Theta)$. For each attribute $f_i \in E(\Theta)$, construct the decision tree $\Gamma(f_i)$ and choose among these trees a tree with the minimum depth. Return this tree as $\Gamma_a^{(k)}(\Theta)$.

We now remind the notion of admissible hypothesis. A hypothesis $H = \{f_1 = \delta_1, \ldots, f_n = \delta_n\}$ over T is called admissible for Θ and an attribute $f_i \in F(T) = \{f_1, \ldots, f_n\}$ if, for any $\sigma \in E(T, f_i) \setminus \{\delta_i\}$, $\Theta\{f_i = \sigma\} \neq \Theta$. The hypothesis H is called admissible for Θ if it is admissible for Θ and any attribute $f_i \in F(T)$.

Let us consider a decision tree $\Gamma(H)$ for Θ in which the root is labeled with an admissible for Θ hypothesis $H = \{f_1 = \delta_1, \ldots, f_n = \delta_n\}$. The set of answers for the query corresponding to the hypothesis H is equal to $A(H) = \{H, \{f_1 = \sigma_1\}, \ldots, \{f_n = \sigma_n\} : \sigma_1 \in E(T, f_1) \setminus \{\delta_1\}, \ldots, \sigma_n \in E(T, f_n) \setminus \{\delta_n\}\}$. For each $S \in A(H)$, there is an edge that leaves the root of $\Gamma(H)$ and enters a node $v(S)$. This edge is labeled with the equation system S. The node $v(S)$ is the root of the decision tree $\Gamma^{(k)}(\Theta S)$.

It is clear that

$$h(\Gamma(H)) = 1 + \max\{h(\Gamma^{(k)}(\Theta S)) : S \in A(H)\} . \tag{4.2}$$

Using (4.2), one can show that $\Gamma(H)$ is a decision tree with the minimum depth for Θ in which the root is labeled with the hypothesis H and which uses for the subtables corresponding to the children of the root decision trees of type k.

We should not consider hypotheses that are not admissible for Θ since, for each such hypothesis H for corresponding query, there is an answer $S \in A(H)$ with $\Theta S = \Theta$, i.e., based on this hypothesis we cannot construct an optimal decision tree for Θ.

Construction of the tree $\Gamma_h^{(k)}(\Theta)$. First, we construct a hypothesis

$$H_\Theta = \{f_1 = \delta_1, \dots, f_n = \delta_n\}$$

for Θ. Let $f_i \in F(T) \setminus E(\Theta)$. Then δ_i is equal to the only number in the set $E(\Theta, f_i)$. Let $f_i \in E(\Theta)$. Then δ_i is the minimum number from $E(\Theta, f_i)$ for which $h(\Gamma^{(k)}(\Theta\{f_i = \delta_i\})) = \max\{h(\Gamma^{(k)}(\Theta\{f_i = \sigma\})) : \sigma \in E(\Theta, f_i)\}$. It is clear that H_Θ is admissible for Θ. Return the tree $\Gamma(H_\Theta)$ as $\Gamma_h^{(k)}(\Theta)$. Using (4.2), one can prove the correctness of this procedure.

Construction of the tree $\Gamma_p^{(k)}(\Theta)$. For each row $r = (\delta_1, \dots, \delta_n)$ of the decision table T, we check if the corresponding proper hypothesis $H_r = \{f_1 = \delta_1, \dots, f_n = \delta_n\}$ is admissible for Θ. For each admissible for Θ proper hypothesis $H_r = \{f_1 = \delta_1, \dots, f_n = \delta_n\}$, we construct the decision tree $\Gamma(H_r)$ and choose among these trees a tree with the minimum depth. Return this tree as $\Gamma_p^{(k)}(\Theta)$.

We now describe an algorithm $\mathscr{C}_h$ that, for a given nonempty decision table T and given $k \in \{1, \dots, 5\}$, constructs for each node Θ of the DAG $\Delta(T)$ a decision tree $\Gamma^{(k)}(\Theta)$, which is a decision tree of type k for the table Θ with the minimum depth.

Algorithm $\mathscr{C}_h$ (construction of the tree $\Gamma^{(k)}(T)$).

Input: A nonempty decision table T, the directed acyclic graph $\Delta(T)$, and a number $k \in \{1, \dots, 5\}$

Output: The decision tree $\Gamma^{(k)}(T)$.

1. If a decision tree is attached to each node of the DAG $\Delta(T)$, then return the tree attached to the node T as $\Gamma^{(k)}(T)$ and halt the algorithm. Otherwise, choose a node Θ of the graph $\Delta(T)$ without attached tree, which is either a terminal node of $\Delta(T)$ or a nonterminal node of $\Delta(T)$ for which all children have attached trees.
2. If Θ is a terminal node, then attach to it the decision tree $\Gamma^{(k)}(\Theta)$ that contains only one node labeled with the decision attached to all rows of Θ and proceed to step 1.
3. If Θ is not a terminal node, then depending on the value k do the following:

   - In the case $k = 1$, construct the tree $\Gamma_a^{(1)}(\Theta)$ and attach it to Θ as the tree $\Gamma^{(1)}(\Theta)$.
   - In the case $k = 2$, construct the tree $\Gamma_h^{(2)}(\Theta)$ and attach it to Θ as the tree $\Gamma^{(2)}(\Theta)$.
   - In the case $k = 3$, construct the trees $\Gamma_a^{(3)}(\Theta)$ and $\Gamma_h^{(3)}(\Theta)$, choose between them a tree with the minimum depth and attach it to Θ as the tree $\Gamma^{(3)}(\Theta)$.
   - In the case $k = 4$, construct the tree $\Gamma_p^{(4)}(\Theta)$ and attach it to Θ as the tree $\Gamma^{(4)}(\Theta)$.
   - In the case $k = 5$, construct the trees $\Gamma_a^{(5)}(\Theta)$ and $\Gamma_p^{(5)}(\Theta)$, choose between them a tree with the minimum depth and attach it to Θ as the tree $\Gamma^{(5)}(\Theta)$.

Proceed to step 1.

Remark 4.1 One can show that the time complexity of the algorithm $\mathcal{C}_h$ is bounded from above by a polynomial on the size of the input table T and the number $|SEP(T)|$ of different separable subtables of T.

Let T be a decision table and $k \in \{1, \ldots, 5\}$. We will use the following notation: $l_h^{(k)}(T) = l(T, \Gamma^{(k)}(T))$ and $c_h^{(k)}(T) = c(T, \Gamma^{(k)}(T))$ – see corresponding definitions in Sect. 2.3.

4.2 Construction of Decision Trees with Minimum Number of Working Nodes

Let T be a nonempty decision table with n conditional attributes $f_1, \ldots, f_n$ and $k \in \{1, \ldots, 5\}$. We can use the DAG $\Delta(T)$ (builded by the algorithm $\mathcal{A}_{DAG}$ described in Sect. 3.1) to construct a decision tree $G^{(k)}(T)$ of type k with the minimum number of working nodes for the decision table T. To construct the tree $G^{(k)}(T)$, for each node Θ of the DAG $\Delta(T)$, we construct a decision tree $G^{(k)}(\Theta)$ of type k with the minimum number of working nodes for the decision table Θ. It will be convenient for us to consider not only subtables that are nodes of $\Delta(T)$ but also empty subtable Λ of T and subtables T_r that contain only one row r of T and are not nodes of $\Delta(T)$. We begin with these special subtables and terminal nodes of $\Delta(T)$ and step-by-step move to the table T.

Let Θ be a terminal node of $\Delta(T)$ or $\Theta = T_r$ for some row r of T. Then $G^{(k)}(\Theta)$ is the decision tree that contains only one node labeled with the decision attached to all rows of Θ. If $\Theta = \Lambda$, then $G^{(k)}(\Theta)$ is the decision tree that contains only one node labeled with 0 (this tree will be considered as a decision tree for Λ).

Let Θ be a nonterminal node of $\Delta(T)$ such that, for each child Θ' of Θ, we already constructed a decision tree $G^{(k)}(\Theta')$. Based on these trees, we can construct a decision tree with the minimum number of working nodes for Θ, which uses for the subtables corresponding to the children of the root decision trees of type k and in which the root is labeled

- With an attribute from $F(T)$ (we denote by $G_a^{(k)}(\Theta)$ such a decision tree).
- With a hypothesis over T (we denote by $G_h^{(k)}(\Theta)$ such a decision tree).
- With a proper hypothesis over T (we denote by $G_p^{(k)}(\Theta)$ such a decision tree).

Since Θ is nondegenerate, the set $E(\Theta)$ is nonempty. We now describe three procedures for construction of the trees $G_a^{(k)}(\Theta)$, $G_h^{(k)}(\Theta)$, and $G_p^{(k)}(\Theta)$, respectively.

Let us consider a decision tree $G(f_i)$ for Θ in which the root is labeled with an attribute $f_i \in E(\Theta)$. For each $\delta \in E(T, f_i)$, there is an edge that leaves the root and enters a node $v(\delta)$. This edge is labeled with the equation system $\{f_i = \delta\}$. The node $v(\delta)$ is the root of the decision tree $G^{(k)}(\Theta\{f_i = \delta\})$.

It is clear that

$$L_w(G(f_i)) = 1 + \sum_{\delta \in E(T, f_i)} L_w(G^{(k)}(\Theta\{f_i = \delta\})) . \tag{4.3}$$

Using (4.3), one can show that $G(f_i)$ is a decision tree with the minimum number of working nodes for Θ in which the root is labeled with the attribute f_i and which uses for the subtables corresponding to the children of the root decision trees of type k.

We should not consider attributes $f_i \in F(T) \setminus E(\Theta)$ since, for each such attribute, there is $\delta \in E(T, f_i)$ with $\Theta\{f_i = \delta\} = \Theta$, i.e., based on this attribute we cannot construct an optimal decision tree for Θ.

Construction of the tree $G_a^{(k)}(\Theta)$. Construct the set of attributes $E(\Theta)$. For each attribute $f_i \in E(\Theta)$, construct the decision tree $G(f_i)$ and choose among these trees a tree with the minimum number of working nodes. Return this tree as $G_a^{(k)}(\Theta)$.

Let us consider a decision tree $G(H)$ for Θ in which the root is labeled with an admissible for Θ hypothesis $H = \{f_1 = \delta_1, \ldots, f_n = \delta_n\}$. The set of answers for the query corresponding to the hypothesis H is equal to $A(H) = \{H, \{f_1 = \sigma_1\}, \ldots, \{f_n = \sigma_n\} : \sigma_1 \in E(T, f_1) \setminus \{\delta_1\}, \ldots, \sigma_n \in E(T, f_n) \setminus \{\delta_n\}\}$. For each $S \in A(H)$, there is an edge that leaves the root of $G(H)$ and enters a node $v(S)$. This edge is labeled with the equation system S. The node $v(S)$ is the root of the decision tree $G^{(k)}(\Theta S)$. It is clear that

$$L_w(G(H)) = 1 + \sum_{S \in A(H)} L_w(G^{(k)}(\Theta S)) . \tag{4.4}$$

Using (4.4), one can show that $G(H)$ is a decision tree with the minimum number of working nodes for Θ in which the root is labeled with the hypothesis H and which uses for the subtables corresponding to the children of the root decision trees of type k.

We should not consider hypotheses that are not admissible for Θ since, for each such hypothesis H for corresponding query, there is an answer $S \in A(H)$ with $\Theta S = \Theta$, i.e., based on this hypothesis we cannot construct an optimal decision tree for Θ.

Construction of the tree $G_h^{(k)}(\Theta)$. First, we construct a hypothesis

$$H_\Theta = \{f_1 = \delta_1, \ldots, f_n = \delta_n\}$$

for Θ. Let $f_i \in F(T) \setminus E(\Theta)$. Then δ_i is equal to the only number in the set $E(\Theta, f_i)$. Let $f_i \in E(\Theta)$. Then δ_i is the minimum number from $E(\Theta, f_i)$ for which $L_w(G^{(k)}(\Theta\{f_i = \delta_i\})) = \max\{L_w(G^{(k)}(\Theta\{f_i = \sigma\})) : \sigma \in E(\Theta, f_i)\}$. It is clear that H_Θ is admissible for Θ. Return the tree $G(H_\Theta)$ as $G_h^{(k)}(\Theta)$. Using (4.4), one can prove the correctness of this procedure.

Construction of the tree $G_p^{(k)}(\Theta)$. For each row $r = (\delta_1, \ldots, \delta_n)$ of the decision table T, we check if the corresponding proper hypothesis $H_r = \{f_1 = \delta_1, \ldots, f_n = \delta_n\}$ is admissible for Θ. For each admissible for Θ proper hypothesis $H_r = \{f_1 = \delta_1, \ldots, f_n = \delta_n\}$, we construct the decision tree $G(H_r)$ and choose among these trees a tree with the minimum number of working nodes. Return this tree as $G_p^{(k)}(\Theta)$.

We now describe an algorithm $\mathcal{C}_{L_w}$ that, for a given nonempty decision table T and given $k \in \{1, \ldots, 5\}$, constructs for each node Θ of the DAG $\Delta(T)$ a decision tree $G^{(k)}(\Theta)$, which is a decision tree of type k for the table Θ with the minimum number of working nodes.

Algorithm $\mathcal{C}_{L_w}$ (construction of the tree $G^{(k)}(T)$).

Input: A nonempty decision table T, the directed acyclic graph $\Delta(T)$, and a number $k \in \{1, \ldots, 5\}$.

Output: The decision tree $G^{(k)}(T)$.

1. If a decision tree is attached to each node of the DAG $\Delta(T)$, then return the tree attached to the node T as $G^{(k)}(T)$ and halt the algorithm. Otherwise, choose a node Θ of the graph $\Delta(T)$ without attached tree, which is either a terminal node of $\Delta(T)$ or a nonterminal node of $\Delta(T)$ for which all children have attached trees.
2. If Θ is a terminal node, then attach to it the decision tree $G^{(k)}(\Theta)$ that contains only one node labeled with the decision attached to all rows of Θ and proceed to step 1.
3. If Θ is not a terminal node, then depending on the value k do the following:

   - In the case $k = 1$, construct the tree $G_a^{(1)}(\Theta)$ and attach it to Θ as the tree $G^{(1)}(\Theta)$.
   - In the case $k = 2$, construct the tree $G_h^{(2)}(\Theta)$ and attach it to Θ as the tree $G^{(2)}(\Theta)$.
   - In the case $k = 3$, construct the trees $G_a^{(3)}(\Theta)$ and $G_h^{(3)}(\Theta)$, choose between them a tree with the minimum number of working nodes and attach it to Θ as the tree $G^{(3)}(\Theta)$.
   - In the case $k = 4$, construct the tree $G_p^{(4)}(\Theta)$ and attach it to Θ as the tree $G^{(4)}(\Theta)$.
   - In the case $k = 5$, construct the trees $G_a^{(5)}(\Theta)$ and $G_p^{(5)}(\Theta)$, choose between them a tree with the minimum number of working nodes and attach it to Θ as the tree $G^{(5)}(\Theta)$.

   Proceed to step 1.

Remark 4.2 One can show that the time complexity of the algorithm $\mathcal{C}_{L_w}$ is bounded from above by a polynomial on the size of the input table T and the number $|SEP(T)|$ of different separable subtables of T.

Let T be a decision table and $k \in \{1, \ldots, 5\}$. We will use the following notation: $l_{L_w}^{(k)}(T) = l(T, G^{(k)}(T))$ and $c_{L_w}^{(k)}(T) = c(T, G^{(k)}(T))$ – see corresponding definitions in Sect. 2.3.

4.3 On Construction of Optimal Decision Trees for L and L_t

First, we discuss a way that allows us to construct optimal decision trees relative to the number of realizable nodes. In Sect. 3.3, we proposed the algorithm $\mathscr{A}_L$ that, for a given decision table T and number $k \in \{1, \ldots, 5\}$, computes the value $L^{(k)}(T)$ – the minimum number of realizable nodes in a decision tree of type k for the table T.

It is difficult to remake this algorithm to construct optimal decision trees. The reason is the following: the considered below procedure "Computation of $L_h^{(k)}(\Theta)$" from Sect. 3.3 is not adapted for the deriving an optimal hypothesis from the whole set of hypotheses.

Let T be a nonempty decision table with n conditional attributes $f_1, \ldots, f_n$ and Θ be a nonterminal node of $\Delta(T)$ such that, for each child Θ' of Θ, we already know the value $L^{(k)}(\Theta')$. Recall that $L_h^{(k)}(\Theta)$ is the minimum number of realizable relative to Θ nodes in a decision tree for Θ, which uses for the subtables corresponding to the children of the root decision trees of type k and in which the root is labeled with a hypothesis over T. The hypotheses on which this minimum is reached will be called optimal.

Computation of $L_h^{(k)}(\Theta)$. For each $f_i \in \{f_1, \ldots, f_n\}$, we compute the value $a_i(\Theta) = \max\{L^{(k)}(\Theta\{f_i = \sigma\}) : \sigma \in E(\Theta, f_i)\}$ and construct set $C(\Theta, f_i) = \{\sigma \in E(\Theta, f_i) : L^{(k)}(\Theta\{f_i = \sigma\}) = a_i(\Theta)\}$. For a tuple $\bar{\delta} = (\delta_1, \ldots, \delta_n) \in C(\Theta) = C(\Theta, f_1) \times \cdots \times C(\Theta, f_n)$, we compute the value

$$K_{\min} = 1 + \sum_{f_i \in E(\Theta), \sigma \in E(\Theta, f_i) \setminus \{\delta_i\}} L^{(k)}(\Theta\{f_i = \sigma\}). \tag{4.5}$$

Then we count the number N of rows from Θ that belong to the set $C(\Theta)$ and compute the cardinality $|C(\Theta)|$ of the set $C(\Theta)$ that is equal to $|C(\Theta, f_1)| \cdot \ldots \cdot |C(\Theta, f_n)|$. As a result, we have $L_h^{(k)}(\Theta) = K_{\min}$ if $N < |C(\Theta)|$ and $L_h^{(k)}(\Theta) = 1 + K_{\min}$ if $N = |C(\Theta)|$.

We can transform this procedure such that it will be possible not only to compute the value $L_h^{(k)}(\Theta)$, but also to find an optimal hypothesis. We denote by $Rows(\Theta)$ the set of rows of the table Θ.

Computation of $L_h^{(k)}(\Theta)$ and finding an optimal hypothesis. For each $f_i \in \{f_1, \ldots, f_n\}$, we compute the value $a_i(\Theta) = \max\{L^{(k)}(\Theta\{f_i = \sigma\}) : \sigma \in E(\Theta, f_i)\}$ and construct the set $C(\Theta, f_i) = \{\sigma \in E(\Theta, f_i) : L^{(k)}(\Theta\{f_i = \sigma\}) = a_i(\Theta)\}$. For a tuple $\bar{\delta} = (\delta_1, \ldots, \delta_n) \in C(\Theta) = C(\Theta, f_1) \times \cdots \times C(\Theta, f_n)$, using (4.5) we compute the value $K_{\min}$. We compute the number $|Rows(\Theta)|$ of rows in Θ and the cardinality $|C(\Theta)|$ of the set $C(\Theta)$ that is equal to $|C(\Theta, f_1)| \cdot \ldots \cdot |C(\Theta, f_n)|$. We consider lexicographical order on the set $C(\Theta)$: $C(\Theta) = \{\bar{b}_1, \ldots, \bar{b}_{|C(\Theta)|}\}$. Let $\bar{b}_i \in C(\Theta)$ and $\bar{b}_i = (\delta_1, \ldots, \delta_n)$. We denote by $H_{\bar{b}_i}$ the hypothesis $\{f_1 = \delta_1, \ldots, f_n = \delta_n\}$.

Set $M = \min(|Rows(\Theta)|, |C(\Theta)|)$. For $i = 1, \ldots, M$, we check if the tuple $\bar{b}_i$ belongs to the set $Rows(\Theta)$. If $\bar{b}_i \notin Rows(\Theta)$ for some $i \in \{1, \ldots, M\}$, then $L_h^{(k)}(\Theta) = K_{\min}$ and $H_{\bar{b}_i}$ is an optimal hypothesis. Let $\bar{b}_i \in Rows(\Theta)$ for $i = 1, \ldots, M$. Then remove all rows $\bar{b}_1, \ldots, \bar{b}_M$ from the set $Rows(\Theta)$ and denote the obtained set A. If $A \neq \emptyset$, then

$L_h^{(k)}(\Theta) = 1 + K_{\min}$ and $H_{\bar{b}_1}$ is an optimal hypothesis. Let $A = \emptyset$. If $|Rows(\Theta)| = |C(\Theta)|$, then $L_h^{(k)}(\Theta) = 1 + K_{\min}$ and $H_{\bar{b}_1}$ is an optimal hypothesis. If $|Rows(\Theta)| < |C(\Theta)|$, then $L_h^{(k)}(\Theta) = K_{\min}$ and $H_{\bar{b}_{M+1}}$ is an optimal hypothesis.

One can show that, based on this new procedure, it is possible to design an algorithm for the construction of decision trees with the minimum number of realizable nodes such that the time complexity of this algorithm will be bounded from above by a polynomial on the size of the input decision table and the number of different separable subtables of it.

Similar approach can be used for the construction of decision trees with the minimum number of realizable terminal nodes.

4.4 Results of Experiments

We make experiments on eight decision tables from the UCI ML Repository [3]. Table 3.1 contains information about each of these decision tables: its name, the number of rows, and the number of attributes. For each of the considered two cost functions, each of the considered five types of decision trees, and each of the considered eight decision tables, we construct an optimal decision tree of the given type for the given table and study decision rules derived from this tree.

For $n = 3, \ldots, 6$, we randomly generate 100 Boolean functions with n variables. For each of the considered two cost functions, each of the considered five types of decision trees, and each of the generated Boolean functions, using its decision table representation, we construct an optimal decision tree of the given type computing this function and study decision rules derived from this tree.

4.4.1 Decision Trees with Minimum Depth

Results of experiments on eight decision tables from [3] and decision trees optimal relative to the depth are represented in Tables 4.1 and 4.2. The first column of Table 4.1 contains the name of the considered decision table T. The last five columns contain values $l_h^{(1)}(T), \ldots, l_h^{(5)}(T)$ (minimum values for each decision table are in bold).

The first column of Table 4.2 contains the name of the considered decision table T. The last five columns contain values $c_h^{(1)}(T), \ldots, c_h^{(5)}(T)$ (maximum values for each decision table are in bold).

Results of experiments on Boolean functions and decision trees optimal relative to the depth are represented in Tables 4.3 and 4.4. The first column of Table 4.3 contains the number n of variables in the considered Boolean functions. The last five columns contain information about values $l_h^{(1)}, \ldots, l_h^{(5)}$ in the format ${}_{\min}Avg_{\max}$ (minimum values of Avg for each n are in bold).

Table 4.1 Experimental results for decision tables from [3]: length of decision rules derived from decision trees with minimum depth

Decision table, T	$l_h^{(1)}(T)$	$l_h^{(2)}(T)$	$l_h^{(3)}(T)$	$l_h^{(4)}(T)$	$l_h^{(5)}(T)$
BALANCE-SCALE	3.60	**3.20**	**3.20**	3.21	**3.20**
BREAST-CANCER	3.61	**2.68**	2.71	2.70	2.78
CARS	3.72	**2.44**	2.48	3.07	3.02
HAYES-ROTH-DATA	2.83	2.22	**2.16**	2.32	2.35
NURSERY	5.78	3.16	4.53	**3.12**	4.50
SOYBEAN-SMALL	1.89	**1.00**	1.89	1.55	1.89
TIC-TAC-TOE	5.09	**3.04**	3.40	3.24	3.14
ZOO-DATA	3.69	**1.56**	2.42	2.17	3.24
Average	3.78	2.41	2.85	2.67	3.01

Table 4.2 Experimental results for decision tables from [3]: coverage of decision rules derived from decision trees with minimum depth

Decision table, T	$c_h^{(1)}(T)$	$c_h^{(2)}(T)$	$c_h^{(3)}(T)$	$c_h^{(4)}(T)$	$c_h^{(5)}(T)$
BALANCE-SCALE	2.60	**4.19**	4.18	4.16	**4.19**
BREAST-BREAST	4.98	9.30	8.36	**9.38**	6.90
CARS	197.60	**332.76**	330.35	97.20	99.42
HAYES-ROTH-DATA	3.46	6.20	**6.49**	5.48	5.45
NURSERY	29.33	1524.04	304.71	**1530.50**	246.14
SOYBEAN-SMALL	3.47	**12.53**	3.47	10.62	3.47
TIC-TAC-TOE	8.38	**66.01**	27.23	56.64	61.45
ZOO-DATA	7.88	10.78	9.80	**10.86**	6.46
Average	32.21	245.73	86.82	215.60	54.18

The first column of Table 4.4 contains the number n of variables in the considered Boolean functions. The last five columns contain information about values $c_h^{(1)}, \ldots, c_h^{(5)}$ in the format ${}_{\min}Avg_{\max}$ (maximum values of Avg for each n are in bold).

4.4.2 Decision Trees with Minimum Number of Working Nodes

Results of experiments on eight decision tables from [3] and decision trees optimal relative to the number of working nodes are represented in Tables 4.5 and 4.6. The first column of Table 4.5 contains the name of the considered decision table T. The last five columns contain values $l_{L_w}^{(1)}(T), \ldots, l_{L_w}^{(5)}(T)$ (minimum values for each decision table are in bold).

Table 4.3 Experimental results for Boolean functions: length of decision rules derived from decision trees with minimum depth

Number of variables, n	$l_h^{(1)}$	$l_h^{(2)}$	$l_h^{(3)}$	$l_h^{(4)}$	$l_h^{(5)}$
3	${}_{1.50}2.20_{2.75}$	${}_{1.25}2.05_{2.63}$	${}_{1.25}\mathbf{1.99}_{2.50}$	${}_{1.25}2.08_{2.63}$	${}_{1.25}2.01_{2.50}$
4	${}_{1.88}3.18_{3.75}$	${}_{1.63}2.92_{3.50}$	${}_{1.63}\mathbf{2.87}_{3.50}$	${}_{1.63}2.91_{3.50}$	${}_{1.63}2.94_{3.50}$
5	${}_{3.44}4.09_{4.63}$	${}_{3.00}3.64_{4.22}$	${}_{2.97}\mathbf{3.60}_{4.06}$	${}_{3.13}3.66_{4.13}$	${}_{3.09}3.70_{4.19}$
6	${}_{4.78}5.14_{5.47}$	${}_{3.98}4.36_{4.77}$	${}_{3.98}4.41_{4.75}$	${}_{3.97}\mathbf{4.35}_{4.70}$	${}_{4.03}4.46_{4.78}$

Table 4.4 Experimental results for Boolean functions: coverage of decision rules derived from decision trees with minimum depth

Number of variables, n	$c_h^{(1)}$	$c_h^{(2)}$	$c_h^{(3)}$	$c_h^{(4)}$	$c_h^{(5)}$
3	${}_{1.25}1.94_{3.00}$	${}_{1.38}\mathbf{2.22}_{3.63}$	${}_{1.50}2.21_{3.63}$	${}_{1.38}2.14_{3.63}$	${}_{1.50}2.17_{3.63}$
4	${}_{1.25}1.99_{5.38}$	${}_{1.50}\mathbf{2.57}_{6.44}$	${}_{1.50}2.52_{6.44}$	${}_{1.50}2.53_{6.44}$	${}_{1.50}2.36_{6.44}$
5	${}_{1.38}2.10_{3.69}$	${}_{2.03}\mathbf{3.03}_{4.56}$	${}_{2.06}2.98_{4.97}$	${}_{2.03}2.93_{4.69}$	${}_{1.81}2.76_{4.66}$
6	${}_{1.59}2.03_{2.84}$	${}_{2.69}\mathbf{3.55}_{4.81}$	${}_{2.58}3.37_{4.64}$	${}_{2.72}3.53_{4.70}$	${}_{2.53}3.24_{4.69}$

Table 4.5 Experimental results for decision tables from [3]: length of decision rules derived from decision trees with minimum number of working nodes

Decision table, T	$l_{L_w}^{(1)}(T)$	$l_{L_w}^{(2)}(T)$	$l_{L_w}^{(3)}(T)$	$l_{L_w}^{(4)}(T)$	$l_{L_w}^{(5)}(T)$
BALANCE-SCALE	3.55	**3.20**	3.53	**3.20**	3.53
BREAST-BREAST	4.98	**2.72**	5.30	2.73	5.27
CARS	2.97	**2.49**	2.96	**2.49**	2.96
HAYES-ROTH-DATA	2.64	**2.22**	2.61	2.23	2.61
NURSERY	3.77	**3.19**	3.77	**3.19**	3.77
SOYBEAN-SMALL	1.34	**1.00**	1.34	1.51	1.34
TIC-TAC-TOE	4.45	3.35	4.41	**3.15**	4.45
ZOO-DATA	3.05	**1.69**	3.05	2.39	3.05
Average	3.34	2.48	3.37	2.61	3.37

The first column of Table 4.6 contains the name of the considered decision table T. The last five columns contain values $c_{L_w}^{(1)}(T), \ldots, c_{L_w}^{(5)}(T)$ (maximum values for each decision table are in bold).

Results of experiments on Boolean functions and decision trees optimal relative to the number of working nodes are represented in Tables 4.7 and 4.8. The first column of Table 4.7 contains the number n of variables in the considered Boolean functions. The last five columns

Table 4.6 Experimental results for decision tables from [3]: coverage of decision rules derived from decision trees with minimum number of working nodes

Decision table, T	$c^{(1)}_{L_w}(T)$	$c^{(2)}_{L_w}(T)$	$c^{(3)}_{L_w}(T)$	$c^{(4)}_{L_w}(T)$	$c^{(5)}_{L_w}(T)$
BALANCE-SCALE	2.79	**4.21**	2.88	**4.21**	2.88
BREAST-BREAST	2.73	8.96	3.05	**9.06**	3.15
CARS	237.33	**332.46**	237.37	**332.46**	237.37
HAYES-ROTH-DATA	3.84	**6.20**	3.87	**6.20**	3.87
NURSERY	1,471.45	**1527.95**	1471.47	**1527.95**	1,471.47
SOYBEAN-SMALL	11.51	**12.53**	11.51	9.81	11.51
TIC-TAC-TOE	22.49	30.19	23.50	**56.50**	22.69
ZOO-DATA	**10.69**	10.68	**10.69**	10.63	**10.69**
Average	220.35	241.65	220.54	244.60	220.45

Table 4.7 Experimental results for Boolean functions: length of decision rules derived from decision trees with minimum number of working nodes

Number of variables, n	$l^{(1)}_{L_w}$	$l^{(2)}_{L_w}$	$l^{(3)}_{L_w}$	$l^{(4)}_{L_w}$	$l^{(5)}_{L_w}$
3	${}_{1.50}2.07_{2.75}$	${}_{1.25}2.06_{2.63}$	${}_{1.25}\mathbf{1.94}_{2.50}$	${}_{1.25}2.06_{2.63}$	${}_{1.25}\mathbf{1.94}_{2.50}$
4	${}_{1.88}2.90_{3.50}$	${}_{1.63}2.94_{3.50}$	${}_{1.81}\mathbf{2.79}_{3.50}$	${}_{1.63}2.94_{3.50}$	${}_{1.81}\mathbf{2.79}_{3.50}$
5	${}_{3.13}3.75_{4.19}$	${}_{3.00}3.77_{4.31}$	${}_{3.13}\mathbf{3.69}_{4.25}$	${}_{3.00}3.77_{4.31}$	${}_{3.13}\mathbf{3.69}_{4.25}$
6	${}_{4.28}4.69_{5.06}$	${}_{4.13}4.69_{5.19}$	${}_{4.25}\mathbf{4.61}_{4.98}$	${}_{4.13}4.69_{5.19}$	${}_{4.25}\mathbf{4.61}_{4.98}$

Table 4.8 Experimental results for Boolean functions: coverage of decision rules derived from decision trees with minimum number of working nodes

Number of variables, n	$c^{(1)}_{L_w}$	$c^{(2)}_{L_w}$	$c^{(3)}_{L_w}$	$c^{(4)}_{L_w}$	$c^{(5)}_{L_w}$
3	${}_{1.25}2.14_{3.00}$	${}_{1.38}2.22_{3.63}$	${}_{1.50}\mathbf{2.27}_{3.63}$	${}_{1.38}2.22_{3.63}$	${}_{1.50}\mathbf{2.27}_{3.63}$
4	${}_{1.63}2.51_{5.38}$	${}_{1.50}2.56_{6.44}$	${}_{1.50}\mathbf{2.63}_{5.44}$	${}_{1.50}2.56_{6.44}$	${}_{1.50}\mathbf{2.63}_{5.44}$
5	${}_{1.88}2.79_{4.19}$	${}_{1.94}\mathbf{2.91}_{4.59}$	${}_{1.75}2.82_{4.34}$	${}_{1.94}\mathbf{2.91}_{4.59}$	${}_{1.75}2.82_{4.34}$
6	${}_{2.16}2.88_{4.09}$	${}_{2.09}\mathbf{3.16}_{4.58}$	${}_{2.27}2.99_{4.11}$	${}_{2.09}\mathbf{3.16}_{4.58}$	${}_{2.27}2.99_{4.11}$

contain information about values $l^{(1)}_{L_w}, \ldots, l^{(5)}_{L_w}$ in the format ${}_{\min}Avg_{\max}$ (minimum values of Avg for each n are in bold).

The first column of Table 4.8 contains the number of variables n in the considered Boolean functions. The last five columns contain information about values $c^{(1)}_{L_w}, \ldots, c^{(5)}_{L_w}$ in the format ${}_{\min}Avg_{\max}$ (maximum values of Avg for each n are in bold).

4.4.3 Analysis of Experimental Results

The results of computer experiments show that the decision rules derived from decision trees with hypotheses (decision trees of types 2, ..., 5) in many cases are better than the ones derived from conventional decision trees (decision trees of type 1). In particular, in the case of decision trees with the minimum depth, for each row in Tables 4.1–4.4, results for the decision trees of type 2 are better than the results for the decision trees of type 1. In the case of decision trees with the minimum number of working nodes, for each row of Tables 4.5–4.8 (with the exception of the row ZOO-DATA in Table 4.6) there is a number $k \in \{2, \ldots, 5\}$ such that the results for the decision trees of type k are better than the results for the decision trees of type 1.

Note that, for the decision trees with the minimum depth, for each decision table from [3] considered in this chapter, the best results related to the length and the coverage among decision trees of types $\{2, \ldots, 5\}$ are close to the optimal ones obtained in [1] with the help of dynamic programming algorithms for the construction of optimal decision rules. Results for the decision trees of type 1 are, generally, far from the optimal ones.

4.5 Conclusions

In this chapter, we designed dynamic programming algorithms for the construction of optimal decision trees relative to two cost functions, constructed such trees for eight datasets from the UCI Machine Learning Repository and randomly generated Boolean functions, and compared the length and coverage of decision rules derived from the constructed decision trees. Experimental results showed that the decision rules derived from the decision trees with hypotheses in many cases are better than the ones derived from the conventional decision trees.

References

1. Amin, T., Chikalov, I., Moshkov, M., Zielosko, B.: Dynamic programming approach for exact decision rule optimization. In: Skowron, A., Suraj, Z. (eds.) Rough Sets and Intelligent Systems—Professor Zdzisław Pawlak in Memoriam—Volume 1. Intelligent Systems Reference Library, vol. 42, pp. 211–228. Springer (2013)
2. Azad, M., Chikalov, I., Hussain, S., Moshkov, M., Zielosko, B.: Decision rules derived from optimal decision trees with hypotheses. Entropy **23**(12), 1641 (2021). https://doi.org/10.3390/e23121641
3. Dua, D., Graff, C.: UCI Machine Learning Repository. University of California, Irvine, School of Information and Computer Sciences (2017). http://archive.ics.uci.edu/ml

5 Greedy Algorithms for Construction of Decision Trees with Hypotheses

There are different ways to construct conventional decision trees including algorithms that can build optimal decision trees for medium-sized decision tables [1–3, 9, 11]. In particular, in [5] (see also Chap. 4), we proposed dynamic programming algorithms for the construction of decision trees with hypotheses that are optimal relative to the depth and relative to the number of working nodes. Such trees often have less complexity than conventional decision trees and can be used as a means for knowledge representation. However, the most common way is to use greedy algorithms [7, 10] that have essentially less time complexity.

In this chapter, which is based on [6], we study the following five types of decision trees: (1) using attributes, (2) using hypotheses, (3) using both attributes and hypotheses, (4) using proper hypotheses, and (5) using both attributes and proper hypotheses. For an arbitrary uncertainty measure, we propose a greedy algorithm that, for a given decision table and type of decision trees, constructs a decision tree of the considered type for this table. This algorithm is a generalization of the entropy-based greedy algorithm proposed in [4].

The first goal of the chapter is to understand which uncertainty measures and types of decision trees should be chosen if we would like to minimize the depth or the number of realizable nodes in the constructed decision trees. To this end, we compare parameters of the decision trees of different types constructed for 10 decision tables from the UCI ML Repository [8] by greedy algorithms based on five uncertainty measures. We do the same for 100 randomly generated Boolean functions with n variables, where $n = 3, \ldots, 6$.

We also study the length and coverage of decision rules derived from the decision trees constructed by greedy algorithms. The second goal of the chapter is to understand which uncertainty measures and types of decision trees should be chosen if we would like to minimize the length or to maximize the coverage of the derived decision rules.

The rest of the chapter is organized as follows. In Sect. 5.1, we consider greedy algorithms for the decision tree construction. Sections 5.2–5.4 contain results of computer experiments and their analysis, and Sect. 5.5—short conclusions.

M. Azad et al., *Decision Trees with Hypotheses*, Synthesis Lectures on Intelligent Technologies, https://doi.org/10.1007/978-3-031-08585-7_5

5.1 Greedy Algorithms

Let U be an uncertainty measure, T be a nondegenerate decision table with n conditional attributes $f_1, \dots, f_n$, and Θ be a nondegenerate subtable of the table T. We now define the *impurity* of a query for the table Θ and uncertainty measure U. The impurity of the query based on an attribute $f_i \in F(T)$ (impurity of query f_i) is equal to $I_U(f_i, \Theta) = \max\{U(\Theta S) : S \in A(f_i)\}$. The impurity of the query based on a hypothesis H (impurity of query H) is equal to $I_U(H, \Theta) = \max\{U(\Theta S) : S \in A(H)\}$.

An attribute f_i is called *admissible* for Θ if it is not constant in Θ. A hypothesis $\{f_1 = \delta_1, \dots, f_n = \delta_n\}$ over T is called admissible for Θ if it satisfies the following condition. For $i = 1, \dots, n$, if f_i is constant in Θ, then δ_i is the only value of f_i in Θ.

We can find by simple search among all attributes an admissible for Θ attribute f_i with the minimum impurity $I_U(f_i, \Theta)$. We can also find by simple search among all proper hypotheses an admissible for Θ proper hypothesis H with the minimum impurity $I_U(H, \Theta)$. It is not necessary to consider all hypotheses to find an admissible for Θ hypothesis with the minimum impurity. For $i = 1, \dots, n$, we denote by δ_i a number from $E(T, f_i)$ such that $U(\Theta\{f_i = \delta_i\}) = \max\{U(\Theta\{f_i = \sigma\}) : \sigma \in E(T, f_i)\}$. Then the hypothesis $H = \{f_1 = \delta_1, \dots, f_n = \delta_n\}$ is admissible for Θ and has the minimum impurity $I_U(H, \Theta)$ among all admissible for Θ hypotheses.

We now describe a greedy algorithm $\mathcal{A}_U$ based on the uncertainty measure U that, for a given nonempty decision table T and $k \in \{1, \dots, 5\}$, constructs a decision tree $\Gamma_U^{(k)}$ of type k for the table T.

Algorithm $\mathcal{A}_U$ (construction of the tree $\Gamma_U^{(k)}$).

Input: A nonempty decision table T and a number $k \in \{1, \dots, 5\}$.
Output: A decision tree of type k for the table T.

1. Construct a tree G consisting of a single node labeled with T.
2. If no node of the tree G is labeled with a table, then the algorithm ends and returns the tree G.
3. Choose a node v in G, which is labeled with a subtable Θ of the table T.
4. If Θ is degenerate, then instead of Θ, we label the node v with 0 if Θ is empty and with the decision attached to each row of Θ if Θ is nonempty.
5. If Θ is nondegenerate, then depending on k we choose an admissible for Θ query X (either attribute or hypothesis) in the following way:

   a. If $k = 1$, then we find an admissible for Θ attribute $X \in F(T)$ with the minimum impurity $I_U(X, \Theta)$.

b. If $k = 2$, then we find an admissible for Θ hypothesis X over T with the minimum impurity $I_U(X, \Theta)$.
c. If $k = 3$, then we find an admissible for Θ attribute $Y \in F(T)$ with the minimum impurity $I_U(Y, \Theta)$ and an admissible for Θ hypothesis Z over T with the minimum impurity $I_U(Z, \Theta)$. Between Y and Z, we choose a query X with the minimum impurity $I_U(X, \Theta)$.
d. If $k = 4$, then we find an admissible for Θ proper hypothesis X over T with the minimum impurity $I_U(X, \Theta)$.
e. If $k = 5$, then we find an admissible for Θ attribute $Y \in F(T)$ with the minimum impurity $I_U(Y, \Theta)$ and an admissible for Θ proper hypothesis Z over T with the minimum impurity $I_U(Z, \Theta)$. Between Y and Z, we choose a query X with the minimum impurity $I_U(X, \Theta)$.

6. Instead of Θ, we label the node v with the query X. For each answer $S \in A(X)$, we add to the tree G a node $v(S)$ and an edge $e(S)$ connecting v and $v(S)$. We label the node $v(S)$ with the subtable ΘS and label the edge $e(S)$ with the answer S. Proceed to step 2.

For a given nonempty decision table T and number $k \in \{1, \ldots, 5\}$, the algorithm $\mathscr{A}_U$ constructs a decision tree $\Gamma_U^{(k)}$ of type k for the table T. We will use the following notation: $h_U^{(k)}(T) = h(\Gamma_U^{(k)})$, $L_U^{(k)}(T) = L(T, \Gamma_U^{(k)})$, $l_U^{(k)}(T) = l(T, \Gamma_U^{(k)})$, $c_U^{(k)}(T) = c(T, \Gamma_U^{(k)})$.

5.2 Results of Experiments on Decision Tables from UCI ML Repository

We make experiments on 10 decision tables from the UCI ML Repository [8]. Table 5.1 contains the information about each of these decision tables: its name, the number of rows, and the number of conditional attributes.

For each decision table and each uncertainty measure $U \in \{me, rme, ent, gini, R\}$, using the algorithm $\mathscr{A}_U$ we construct a decision tree of type k for this table, $k = 1, \ldots, 5$. We evaluate the depth and the number of realizable nodes for this tree, derive decision rules from it, and study them.

5.2.1 Results for Misclassification Error *me*

Results of experiments for the decision tables from [8] and decision trees are represented in Tables 5.2 and 5.3. The first column of Table 5.2 contains the name of the considered decision table T. The last five columns contain values $h_{me}^{(1)}(T), \ldots, h_{me}^{(5)}(T)$ (minimum values for each decision table are in bold).

Table 5.1 Decision tables from [8] used in experiments

Decision table	Number of rows	Number of attributes
BALANCE-SCALE	625	5
BREAST-CANCER	266	10
CARS	1728	7
HAYES-ROTH-DATA	69	5
LYMPHOGRAPHY	148	18
NURSERY	12,960	9
SOYBEAN-SMALL	47	36
SPECT-TEST	169	22
TIC-TAC-TOE	958	10
ZOO-DATA	59	17

Table 5.2 Experimental results for decision tables from [8] (me and h)

Decision table, T	$h_{me}^{(1)}(T)$	$h_{me}^{(2)}(T)$	$h_{me}^{(3)}(T)$	$h_{me}^{(4)}(T)$	$h_{me}^{(5)}(T)$
BALANCE-SCALE	**4**	**4**	**4**	**4**	**4**
BREAST-CANCER	7	6	**5**	6	**5**
CARS	**6**	**6**	**6**	**6**	**6**
HAYES-ROTH-DATA	**4**	**4**	**4**	**4**	**4**
LYMPHOGRAPHY	7	**5**	**5**	7	**5**
NURSERY	**8**	**8**	**8**	**8**	**8**
SOYBEAN-SMALL	**2**	5	**2**	7	**2**
SPECT-TEST	15	**4**	**4**	**4**	**4**
TIC-TAC-TOE	**7**	**7**	**7**	8	**7**
ZOO-DATA	**4**	**4**	**4**	6	**4**
Average	6.4	5.3	4.9	6	4.9

The first column of Table 5.3 contains the name of the considered decision table T. The last five columns contain values $L_{me}^{(1)}(T), \ldots, L_{me}^{(5)}(T)$ (minimum values for each decision table are in bold).

Results of experiments for the decision tables from [8] and decision rules are represented in Tables 5.4 and 5.5. The first column of Table 5.4 contains the name of the considered decision table T. The last five columns contain values $l_{me}^{(1)}(T), \ldots, l_{me}^{(5)}(T)$ (minimum values for each decision table are in bold).

The first column of Table 5.5 contains the name of the considered decision table T. The last five columns contain values $c_{me}^{(1)}(T), \ldots, c_{me}^{(5)}(T)$ (maximum values for each decision table are in bold).

Table 5.3 Experimental results for decision tables from [8] (me and L)

Decision table, T	$L_{me}^{(1)}(T)$	$L_{me}^{(2)}(T)$	$L_{me}^{(3)}(T)$	$L_{me}^{(4)}(T)$	$L_{me}^{(5)}(T)$
BALANCE-SCALE	**556**	5234	3694	5234	3694
BREAST-CANCER	**238**	21,922	285	33,099	254
CARS	**1462**	66,593	1590	66,593	1590
HAYES-ROTH-DATA	**65**	353	88	348	88
LYMPHOGRAPHY	**92**	49,780	126	165,481	126
NURSERY	**4623**	13,487,465	5473	13,487,465	5473
SOYBEAN-SMALL	**23**	5375	**23**	44,601	**23**
SPECT-TEST	**91**	3266	168	6229	168
TIC-TAC-TOE	**547**	287,504	61,315	703,982	613
ZOO-DATA	**27**	1482	**27**	4411	**27**
Average	772.4	1,392,897.4	7278.9	1,451,744.3	1205.6

Table 5.4 Experimental results for decision tables from [8] (me and l)

Decision table, T	$l_{me}^{(1)}(T)$	$l_{me}^{(2)}(T)$	$l_{me}^{(3)}(T)$	$l_{me}^{(4)}(T)$	$l_{me}^{(5)}(T)$
BALANCE-SCALE	3.64	**3.20**	3.24	**3.20**	3.24
BREAST-CANCER	3.61	**2.76**	3.48	2.79	3.55
CARS	5.05	**2.47**	4.99	**2.47**	4.99
HAYES-ROTH-DATA	2.87	2.26	2.86	**2.25**	2.86
LYMPHOGRAPHY	3.48	**1.99**	3.53	2.14	3.53
NURSERY	4.82	**3.34**	4.69	**3.34**	4.69
SOYBEAN-SMALL	1.89	**1.00**	1.89	1.57	1.89
SPECT-TEST	5.44	2.27	4.81	**2.05**	4.81
TIC-TAC-TOE	5.48	3.35	3.36	**3.19**	5.47
ZOO-DATA	2.53	**1.56**	2.53	1.85	2.53
Average	3.88	2.42	3.54	2.48	3.76

5.2.2 Results for Relative Misclassification Error *rme*

Results of experiments for the decision tables from [8] and decision trees are represented in Tables 5.6 and 5.7. The first column of Table 5.6 contains the name of the considered decision table T. The last five columns contain values $h_{rme}^{(1)}(T), \ldots, h_{rme}^{(5)}(T)$ (minimum values for each decision table are in bold).

The first column of Table 5.7 contains the name of the considered decision table T. The last five columns contain values $L_{rme}^{(1)}(T), \ldots, L_{rme}^{(5)}(T)$ (minimum values for each decision table are in bold).

Table 5.5 Experimental results for decision tables from [8] (me and c)

Decision table, T	$c^{(1)}_{me}(T)$	$c^{(2)}_{me}(T)$	$c^{(3)}_{me}(T)$	$c^{(4)}_{me}(T)$	$c^{(5)}_{me}(T)$
BALANCE-SCALE	2.44	**4.21**	4.05	**4.21**	4.05
BREAST-CANCER	4.72	8.38	5.16	**8.84**	4.97
CARS	5.74	**332.63**	5.97	**332.63**	5.97
HAYES-ROTH-DATA	3.58	**6.19**	3.62	**6.19**	3.62
LYMPHOGRAPHY	4.14	18.19	4.51	**18.70**	4.51
NURSERY	300.54	**1523.16**	304.06	**1523.16**	304.06
SOYBEAN-SMALL	3.47	**12.32**	3.47	9.32	3.47
SPECT-TEST	8.75	**57.98**	18.69	57.87	18.69
TIC-TAC-TOE	6.31	49.52	50.26	**56.26**	6.70
ZOO-DATA	6.83	10.59	6.83	**10.80**	6.83
Average	34.65	202.32	40.66	202.80	36.29

Table 5.6 Experimental results for decision tables from [8] (rme and h)

Decision table, T	$h^{(1)}_{rme}(T)$	$h^{(2)}_{rme}(T)$	$h^{(3)}_{rme}(T)$	$h^{(4)}_{rme}(T)$	$h^{(5)}_{rme}(T)$
BALANCE-SCALE	**4**	**4**	**4**	**4**	**4**
BREAST-CANCER	9	9	9	**8**	9
CARS	**6**	**6**	**6**	**6**	**6**
HAYES-ROTH-DATA	**4**	**4**	**4**	**4**	**4**
LYMPHOGRAPHY	**10**	11	**10**	12	**10**
NURSERY	**8**	**8**	**8**	**8**	**8**
SOYBEAN-SMALL	**2**	6	**2**	8	**2**
SPECT-TEST	20	**5**	**5**	14	11
TIC-TAC-TOE	**7**	8	**7**	8	**7**
ZOO-DATA	8	7	**6**	10	**6**
Average	7.8	6.8	6.1	8.2	6.7

Results of experiments for the decision tables from [8] and decision rules are represented in Tables 5.8 and 5.9. The first column of Table 5.8 contains the name of the considered decision table T. The last five columns contain values $l^{(1)}_{rme}(T), \ldots, l^{(5)}_{rme}(T)$ (minimum values for each decision table are in bold).

The first column of Table 5.9 contains the name of the considered decision table T. The last five columns contain values $c^{(1)}_{rme}(T), \ldots, c^{(5)}_{rme}(T)$ (maximum values for each decision table are in bold).

Table 5.7 Experimental results for decision tables from [8] (rme and L)

Decision table, T	$L^{(1)}_{rme}(T)$	$L^{(2)}_{rme}(T)$	$L^{(3)}_{rme}(T)$	$L^{(4)}_{rme}(T)$	$L^{(5)}_{rme}(T)$
BALANCE-SCALE	**556**	5234	3694	5234	3694
BREAST-CANCER	**255**	446,170	304	103,642	266
CARS	**1592**	66,593	1831	66,593	1831
HAYES-ROTH-DATA	73	428	**72**	378	**72**
LYMPHOGRAPHY	**116**	2,475,650	143	1,952,599	143
NURSERY	**4493**	13,487,465	13,667	13,487,465	13,667
SOYBEAN-SMALL	**7**	11,403	**7**	113,855	**7**
SPECT-TEST	**123**	6983	5495	398,926	1116
TIC-TAC-TOE	**648**	864,578	200,847	946,858	940
ZOO-DATA	**35**	6536	37	30,889	37
Average	789.8	1,737,104	22,609.7	1,710,643.9	2177.3

Table 5.8 Experimental results for decision tables from [8] (rme and l)

Decision table, T	$l^{(1)}_{rme}(T)$	$l^{(2)}_{rme}(T)$	$l^{(3)}_{rme}(T)$	$l^{(4)}_{rme}(T)$	$l^{(5)}_{rme}(T)$
BALANCE-SCALE	3.64	**3.20**	3.24	**3.20**	3.24
BREAST-CANCER	6.11	**2.68**	6.01	2.73	6.06
CARS	5.30	**2.47**	5.18	**2.47**	5.18
HAYES-ROTH-DATA	3.22	**2.20**	3.20	2.22	3.20
LYMPHOGRAPHY	6.97	**2.01**	6.80	2.12	6.80
NURSERY	4.72	**3.34**	4.33	**3.34**	4.33
SOYBEAN-SMALL	1.34	**1.00**	1.34	1.53	1.34
SPECT-TEST	8.37	2.28	2.18	**1.79**	2.21
TIC-TAC-TOE	5.77	3.52	3.70	**3.19**	5.63
ZOO-DATA	4.90	**1.66**	4.61	1.83	4.61
Average	5.03	2.43	4.06	2.44	4.26

5.2.3 Results for Entropy *ent*

Results of experiments for the decision tables from [8] and decision trees are represented in Tables 5.10 and 5.11. The first column of Table 5.10 contains the name of the considered decision table T. The last five columns contain values $h^{(1)}_{ent}(T), \ldots, h^{(5)}_{ent}(T)$ (minimum values for each decision table are in bold).

The first column of Table 5.11 contains the name of the considered decision table T. The last five columns contain values $L^{(1)}_{ent}(T), \ldots, L^{(5)}_{ent}(T)$ (minimum values for each decision table are in bold).

Table 5.9 Experimental results for decision tables from [8] (rme and c)

Decision table, T	$c^{(1)}_{rme}(T)$	$c^{(2)}_{rme}(T)$	$c^{(3)}_{rme}(T)$	$c^{(4)}_{rme}(T)$	$c^{(5)}_{rme}(T)$
BALANCE-SCALE	2.44	**4.21**	4.05	**4.21**	4.05
BREAST-CANCER	2.18	9.12	2.68	**9.46**	2.37
CARS	3.95	**332.63**	5.40	**332.63**	5.40
HAYES-ROTH-DATA	1.81	**6.23**	1.81	6.22	1.81
LYMPHOGRAPHY	3.32	19.76	4.12	**21.33**	4.12
NURSERY	1451.06	**1523.16**	1460.80	**1523.16**	1460.80
SOYBEAN-SMALL	11.51	**12.32**	11.51	11.11	11.51
SPECT-TEST	5.54	57.98	57.98	**58.01**	54.76
TIC-TAC-TOE	5.31	50.48	34.99	**56.25**	6.27
ZOO-DATA	7.10	10.69	7.71	**10.97**	7.71
Average	149.42	202.66	159.10	203.33	155.88

Table 5.10 Experimental results for decision tables from [8] (ent and h)

Decision table, T	$h^{(1)}_{ent}(T)$	$h^{(2)}_{ent}(T)$	$h^{(3)}_{ent}(T)$	$h^{(4)}_{ent}(T)$	$h^{(5)}_{ent}(T)$
BALANCE-SCALE	**4**	**4**	**4**	**4**	**4**
BREAST-CANCER	9	9	9	**8**	9
CARS	**6**	**6**	**6**	**6**	**6**
HAYES-ROTH-DATA	**4**	**4**	**4**	**4**	**4**
LYMPHOGRAPHY	11	11	**9**	13	10
NURSERY	**8**	**8**	**8**	**8**	**8**
SOYBEAN-SMALL	**2**	6	**2**	8	**2**
SPECT-TEST	20	**5**	**5**	14	11
TIC-TAC-TOE	**7**	8	**7**	8	**7**
ZOO-DATA	8	6	**5**	8	**5**
Average	7.9	6.7	5.9	8.1	6.6

Results of experiments for the decision tables from [8] and decision rules are represented in Tables 5.12 and 5.13. The first column of Table 5.12 contains the name of the considered decision table T. The last five columns contain values $l^{(1)}_{ent}(T), \ldots, l^{(5)}_{ent}(T)$ (minimum values for each decision table are in bold).

The first column of Table 5.13 contains the name of the considered decision table T. The last five columns contain values $c^{(1)}_{ent}(T), \ldots, c^{(5)}_{ent}(T)$ (maximum values for each decision table are in bold).

Table 5.11 Experimental results for decision tables from [8] (*ent* and *L*)

Decision table, T	$L_{ent}^{(1)}(T)$	$L_{ent}^{(2)}(T)$	$L_{ent}^{(3)}(T)$	$L_{ent}^{(4)}(T)$	$L_{ent}^{(5)}(T)$
BALANCE-SCALE	**556**	5234	4102	5234	4102
BREAST-CANCER	**255**	446,170	304	103,642	266
CARS	**1136**	65,624	3944	65,624	3944
HAYES-ROTH-DATA	73	421	**72**	367	**72**
LYMPHOGRAPHY	**123**	6,653,366	162	8,515,841	153
NURSERY	**4460**	12,790,306	14,422	12,790,306	14,422
SOYBEAN-SMALL	**7**	10,029	**7**	157,640	**7**
SPECT-TEST	**123**	6983	5495	398,926	1116
TIC-TAC-TOE	**648**	864,578	200,847	946,858	940
ZOO-DATA	**33**	2134	35	13,310	35
Average	741.4	2,084,484.5	22,939	2,299,774.8	2505.7

Table 5.12 Experimental results for decision tables from [8] (*ent* and *l*)

Decision table, T	$l_{ent}^{(1)}(T)$	$l_{ent}^{(2)}(T)$	$l_{ent}^{(3)}(T)$	$l_{ent}^{(4)}(T)$	$l_{ent}^{(5)}(T)$
BALANCE-SCALE	3.64	**3.20**	3.24	**3.20**	3.24
BREAST-CANCER	6.11	**2.68**	6.01	2.73	6.06
CARS	4.68	**2.45**	4.17	**2.45**	4.17
HAYES-ROTH-DATA	3.22	**2.17**	3.20	2.22	3.20
LYMPHOGRAPHY	6.61	**1.99**	6.32	2.11	6.37
NURSERY	4.72	**3.80**	4.29	**3.80**	4.29
SOYBEAN-SMALL	1.34	**1.00**	1.34	1.47	1.34
SPECT-TEST	8.37	2.28	2.18	**1.79**	2.21
TIC-TAC-TOE	5.77	3.52	3.70	**3.19**	5.63
ZOO-DATA	4.27	**1.66**	3.98	1.97	3.98
Average	4.87	2.48	3.84	2.49	4.05

5.2.4 Results for Gini Index *gini*

Results of experiments for the decision tables from [8] and decision trees are represented in Tables 5.14 and 5.15. The first column of Table 5.14 contains the name of the considered decision table T. The last five columns contain values $h_{gini}^{(1)}(T), \ldots, h_{gini}^{(5)}(T)$ (minimum values for each decision table are in bold).

The first column of Table 5.15 contains the name of the considered decision table T. The last five columns contain values $L_{gini}^{(1)}(T), \ldots, L_{gini}^{(5)}(T)$ (minimum values for each decision table are in bold).

Table 5.13 Experimental results for decision tables from [8] (ent and c)

Decision table, T	$c^{(1)}_{ent}(T)$	$c^{(2)}_{ent}(T)$	$c^{(3)}_{ent}(T)$	$c^{(4)}_{ent}(T)$	$c^{(5)}_{ent}(T)$
BALANCE-SCALE	2.44	**4.21**	4.03	**4.21**	4.03
BREAST-CANCER	2.18	9.12	2.68	**9.46**	2.37
CARS	15.97	**332.71**	37.12	**332.71**	37.12
HAYES-ROTH-DATA	1.81	**6.38**	1.81	6.26	1.81
LYMPHOGRAPHY	3.51	**21.54**	4.72	21.53	4.42
NURSERY	1451.07	**1516.04**	1462.03	**1516.04**	1462.03
SOYBEAN-SMALL	11.51	**12.32**	11.51	11.66	11.51
SPECT-TEST	5.54	57.98	57.98	**58.01**	54.76
TIC-TAC-TOE	5.31	50.48	34.99	**56.25**	6.27
ZOO-DATA	7.24	**10.66**	7.85	10.36	7.85
Average	150.66	202.14	162.47	202.65	159.22

Table 5.14 Experimental results for decision tables from [8] ($gini$ and h)

Decision table, T	$h^{(1)}_{gini}(T)$	$h^{(2)}_{gini}(T)$	$h^{(3)}_{gini}(T)$	$h^{(4)}_{gini}(T)$	$h^{(5)}_{gini}(T)$
BALANCE-SCALE	**4**	**4**	**4**	**4**	**4**
BREAST-CANCER	9	9	9	**8**	9
CARS	**6**	**6**	**6**	**6**	**6**
HAYES-ROTH-DATA	**4**	**4**	**4**	**4**	**4**
LYMPHOGRAPHY	**10**	11	**10**	13	**10**
NURSERY	**8**	**8**	**8**	**8**	**8**
SOYBEAN-SMALL	**2**	6	**2**	8	**2**
SPECT-TEST	20	**5**	**5**	14	11
TIC-TAC-TOE	**7**	8	**7**	8	**7**
ZOO-DATA	8	**7**	8	9	**7**
Average	7.8	6.8	6.3	8.2	6.8

Results of experiments for the decision tables from [8] and decision rules are represented in Tables 5.16 and 5.17. The first column of Table 5.16 contains the name of the considered decision table T. The last five columns contain values $l^{(1)}_{gini}(T), \ldots, l^{(5)}_{gini}(T)$ (minimum values for each decision table are in bold).

The first column of Table 5.17 contains the name of the considered decision table T. The last five columns contain values $c^{(1)}_{gini}(T), \ldots, c^{(5)}_{gini}(T)$ (maximum values for each decision table are in bold).

Table 5.15 Experimental results for decision tables from [8] ($gini$ and L)

Decision table, T	$L^{(1)}_{gini}(T)$	$L^{(2)}_{gini}(T)$	$L^{(3)}_{gini}(T)$	$L^{(4)}_{gini}(T)$	$L^{(5)}_{gini}(T)$
BALANCE-SCALE	**556**	5234	4006	5234	4006
BREAST-CANCER	**255**	446,170	304	103,642	266
CARS	**1511**	65,579	3643	65,579	3643
HAYES-ROTH-DATA	73	415	**72**	363	**72**
LYMPHOGRAPHY	**115**	2,735,180	140	2,232,312	**115**
NURSERY	4284	12,795,294	16,566	12,795,294	16,566
SOYBEAN-SMALL	**7**	10,029	**7**	148,891	**7**
SPECT-TEST	**123**	6983	5495	398,926	1116
TIC-TAC-TOE	**648**	864,578	200,847	946,858	940
ZOO-DATA	**39**	4702	57	20,896	41
Average	761.1	1,693,416.4	23,113.7	1,671,799.5	2677.2

Table 5.16 Experimental results for decision tables from [8] ($gini$ and l)

Decision table, T	$l^{(1)}_{gini}(T)$	$l^{(2)}_{gini}(T)$	$l^{(3)}_{gini}(T)$	$l^{(4)}_{gini}(T)$	$l^{(5)}_{gini}(T)$
BALANCE-SCALE	3.64	**3.20**	3.23	**3.20**	3.23
BREAST-CANCER	6.11	**2.68**	6.01	2.73	6.06
CARS	5.25	**2.45**	4.70	**2.45**	4.70
HAYES-ROTH-DATA	3.22	**2.20**	3.20	2.23	3.20
LYMPHOGRAPHY	7.01	**1.99**	6.91	2.14	7.01
NURSERY	4.63	**3.76**	4.12	**3.76**	4.12
SOYBEAN-SMALL	1.34	**1.00**	1.34	1.47	1.34
SPECT-TEST	8.37	2.28	2.18	**1.79**	2.21
TIC-TAC-TOE	5.77	3.52	3.70	**3.19**	5.63
ZOO-DATA	5.10	**1.66**	4.71	2.05	4.81
Average	5.04	2.47	4.01	2.50	4.23

5.2.5 Results for Uncertainty Measure R

Results of experiments for the decision tables from [8] and decision trees are represented in Tables 5.18 and 5.19. The first column of Table 5.18 contains the name of the considered decision table T. The last five columns contain values $h^{(1)}_R(T), \ldots, h^{(5)}_R(T)$ (minimum values for each decision table are in bold).

Table 5.17 Experimental results for decision tables from [8] ($gini$ and c)

Decision table, T	$c_{gini}^{(1)}(T)$	$c_{gini}^{(2)}(T)$	$c_{gini}^{(3)}(T)$	$c_{gini}^{(4)}(T)$	$c_{gini}^{(5)}(T)$
BALANCE-SCALE	2.44	**4.21**	4.08	**4.21**	4.08
BREAST-CANCER	2.18	9.12	2.68	**9.46**	2.37
CARS	4.41	**332.70**	26.27	**332.70**	26.27
HAYES-ROTH-DATA	1.81	**6.23**	1.81	6.19	1.81
LYMPHOGRAPHY	3.28	20.85	3.49	**21.45**	3.28
NURSERY	1454.14	**1516.81**	1467.10	**1516.81**	1467.10
SOYBEAN-SMALL	11.51	**12.32**	11.51	11.66	11.51
SPECT-TEST	5.54	57.98	57.98	**58.01**	54.76
TIC-TAC-TOE	5.31	50.48	34.99	**56.25**	6.27
ZOO-DATA	5.85	10.66	7.37	**10.86**	6.46
Average	149.65	202.14	161.73	202.76	158.39

Table 5.18 Experimental results for decision tables from [8] (R and h)

Decision table, T	$h_R^{(1)}(T)$	$h_R^{(2)}(T)$	$h_R^{(3)}(T)$	$h_R^{(4)}(T)$	$h_R^{(5)}(T)$
BALANCE-SCALE	**4**	**4**	**4**	**4**	**4**
BREAST-CANCER	6	6	**5**	6	**5**
CARS	**6**	**6**	**6**	**6**	**6**
HAYES-ROTH-DATA	**4**	**4**	**4**	**4**	**4**
LYMPHOGRAPHY	**5**	6	**5**	7	**5**
NURSERY	8	8	**7**	8	**7**
SOYBEAN-SMALL	**2**	5	**2**	7	**2**
SPECT-TEST	9	**4**	5	6	5
TIC-TAC-TOE	7	**6**	**6**	8	7
ZOO-DATA	**4**	5	**4**	6	**4**
Average	5.5	5.4	4.8	6.2	4.9

The first column of Table 5.19 contains the name of the considered decision table T. The last five columns contain values $L_R^{(1)}(T), \ldots, L_R^{(5)}(T)$ (minimum values for each decision table are in bold).

Results of experiments for the decision tables from [8] and decision rules are represented in Tables 5.20 and 5.21. The first column of Table 5.20 contains the name of the considered decision table T. The last five columns contain values $l_R^{(1)}(T), \ldots, l_R^{(5)}(T)$ (minimum values for each decision table are in bold).

Table 5.19 Experimental results for decision tables from [8] (R and L)

Decision table, T	$L_R^{(1)}(T)$	$L_R^{(2)}(T)$	$L_R^{(3)}(T)$	$L_R^{(4)}(T)$	$L_R^{(5)}(T)$
BALANCE-SCALE	**556**	5234	4006	5234	4006
BREAST-CANCER	**222**	19,698	320	29,308	264
CARS	**1405**	65,579	1541	65,579	1541
HAYES-ROTH-DATA	**55**	338	65	353	65
LYMPHOGRAPHY	**90**	47,278	110	141,322	110
NURSERY	4366	12,795,294	**3889**	12,795,294	**3889**
SOYBEAN-SMALL	**23**	5335	**23**	36,733	**23**
SPECT-TEST	**53**	2730	2126	5920	656
TIC-TAC-TOE	**579**	164,663	74,567	645,542	734
ZOO-DATA	**25**	1542	76	3826	72
Average	737.4	1,310,769.1	8672.3	1,372,911.1	1136

Table 5.20 Experimental results for decision tables from [8] (R and l)

Decision table, T	$l_R^{(1)}(T)$	$l_R^{(2)}(T)$	$l_R^{(3)}(T)$	$l_R^{(4)}(T)$	$l_R^{(5)}(T)$
BALANCE-SCALE	3.64	**3.20**	3.23	**3.20**	3.23
BREAST-CANCER	3.36	2.80	3.25	**2.79**	3.33
CARS	4.91	**2.45**	4.82	**2.45**	4.82
HAYES-ROTH-DATA	2.68	2.26	2.55	**2.23**	2.55
LYMPHOGRAPHY	2.92	**2.01**	3.02	2.14	3.02
NURSERY	5.34	**3.76**	5.29	**3.76**	5.29
SOYBEAN-SMALL	1.89	**1.00**	1.89	1.68	1.89
SPECT-TEST	3.93	2.24	1.88	2.09	**1.84**
TIC-TAC-TOE	5.16	3.44	**3.43**	3.44	5.16
ZOO-DATA	2.41	**1.78**	2.53	2.17	2.53
Average	3.62	2.49	3.19	2.60	3.37

The first column of Table 5.21 contains the name of the considered decision table T. The last five columns contain values $c_R^{(1)}(T), \ldots, c_R^{(5)}(T)$ (maximum values for each decision table are in bold).

Table 5.21 Experimental results for decision tables from [8] (R and c)

Decision table, T	$c_R^{(1)}(T)$	$c_R^{(2)}(T)$	$c_R^{(3)}(T)$	$c_R^{(4)}(T)$	$c_R^{(5)}(T)$
BALANCE-SCALE	2.44	**4.21**	4.08	**4.21**	4.08
BREAST-CANCER	4.62	7.56	4.74	**8.14**	4.81
CARS	7.64	**332.70**	8.00	**332.70**	8.00
HAYES-ROTH-DATA	3.90	**6.19**	3.93	**6.19**	3.93
LYMPHOGRAPHY	4.53	14.91	4.72	**18.11**	4.72
NURSERY	36.52	**1516.81**	36.77	**1516.81**	36.77
SOYBEAN-SMALL	3.47	**12.32**	3.47	10.04	3.47
SPECT-TEST	20.46	55.73	55.73	56.72	**56.75**
TIC-TAC-TOE	12.86	25.19	25.77	**34.12**	13.10
ZOO-DATA	6.86	10.54	6.49	**10.73**	6.49
Average	10.33	198.62	15.37	199.78	14.21

5.3 Results of Experiments on Randomly Generated Boolean Functions

For $n = 3, \ldots, 6$, we randomly generate 100 Boolean functions with n variables. We represent each Boolean function f with n variables as a decision table T_f with n columns labeled with these variables as attributes and with 2^n rows that are all possible n-tuples of values of the variables. Each row is labeled with the decision that is the value of the function f on the corresponding n-tuple. Decision trees for the decision table T_f can be considered as decision trees computing the function f.

For each function f (really for its decision table representation T_f) and each uncertainty measure $U \in \{me, rme, ent, gini, R\}$, using the algorithm $\mathscr{A}_U$ we construct a decision tree of type k computing this function, $k = 1, \ldots, 5$. We evaluate the depth and the number of realizable nodes for this tree, derive decision rules from it, and study them.

For each Boolean function, each hypothesis over the decision table representing it is proper. Therefore, for each Boolean function f and each uncertainty measure U, $h_U^{(2)}(T_f) = h_U^{(4)}(T_f)$, $h_U^{(3)}(T_f) = h_U^{(5)}(T_f)$, $L_U^{(2)}(T_f) = L_U^{(4)}(T_f)$, $L_U^{(3)}(T_f) = L_U^{(5)}(T_f)$, $l_U^{(2)}(T_f) = l_U^{(4)}(T_f)$, $l_U^{(3)}(T_f) = l_U^{(5)}(T_f)$, $c_U^{(2)}(T_f) = c_U^{(4)}(T_f)$, and $c_U^{(3)}(T_f) = c_U^{(5)}(T_f)$.

Results of experiments on randomly generated Boolean functions do not depend on the uncertainty measure $U \in \{me, rme, ent, gini, R\}$. At the end of this section, we will explain this interesting fact.

Results of experiments for Boolean functions and decision trees are represented in Tables 5.22 and 5.23. The first column in Table 5.22 contains the number of variables n in the considered Boolean functions. The last five columns contain information about values

Table 5.22 Experimental results for Boolean functions ($U \in \{me, rme, ent, gini, R\}$ and h)

Number of variables, n	$h_U^{(1)}$	$h_U^{(2)}$	$h_U^{(3)}$	$h_U^{(4)}$	$h_U^{(5)}$
3	$_{0}2.94_{3}$	$_{0}2.02_{3}$	$_{0}\mathbf{1.86}_{3}$	$_{0}2.02_{3}$	$_{0}\mathbf{1.86}_{3}$
4	$_{4}4.00_{4}$	$_{2}3.05_{4}$	$_{2}\mathbf{2.97}_{3}$	$_{2}3.05_{4}$	$_{2}\mathbf{2.97}_{3}$
5	$_{5}5.00_{5}$	$_{4}4.11_{5}$	$_{3}\mathbf{3.99}_{4}$	$_{4}4.11_{5}$	$_{3}\mathbf{3.99}_{4}$
6	$_{6}6.00_{6}$	$_{5}5.09_{6}$	$_{5}\mathbf{5.00}_{5}$	$_{5}5.09_{6}$	$_{5}\mathbf{5.00}_{5}$

Table 5.23 Experimental results for Boolean functions ($U \in \{me, rme, ent, gini, R\}$ and L)

Number of variables, n	$L_U^{(1)}$	$L_U^{(2)}$	$L_U^{(3)}$	$L_U^{(4)}$	$L_U^{(5)}$
3	$_{1}\mathbf{9.60}_{15}$	$_{1}12.33_{22}$	$_{1}9.61_{15}$	$_{1}12.33_{22}$	$_{1}9.61_{15}$
4	$_{15}\mathbf{21.02}_{29}$	$_{14}44.75_{70}$	$_{11}27.69_{58}$	$_{14}44.75_{70}$	$_{11}27.69_{58}$
5	$_{31}\mathbf{42.54}_{51}$	$_{125}218.05_{292}$	$_{25}70.19_{176}$	$_{125}218.05_{292}$	$_{25}70.19_{176}$
6	$_{69}\mathbf{86.30}_{101}$	$_{649}1171.03_{1538}$	$_{75}292.99_{807}$	$_{649}1171.03_{1538}$	$_{75}292.99_{807}$

Table 5.24 Experimental results for Boolean functions ($U \in \{me, rme, ent, gini, R\}$ and l)

Number of variables, n	$l_U^{(1)}$	$l_U^{(2)}$	$l_U^{(3)}$	$l_U^{(4)}$	$l_U^{(5)}$
3	$_{1.75}2.28_{3.00}$	$_{1.25}\mathbf{2.09}_{2.75}$	$_{1.25}2.10_{3.00}$	$_{1.25}\mathbf{2.09}_{2.75}$	$_{1.25}2.10_{3.00}$
4	$_{2.38}3.31_{3.88}$	$_{1.63}\mathbf{2.96}_{3.69}$	$_{1.63}3.05_{3.50}$	$_{1.63}\mathbf{2.96}_{3.69}$	$_{1.63}3.05_{3.50}$
5	$_{3.75}4.29_{4.63}$	$_{3.25}\mathbf{3.92}_{4.50}$	$_{3.03}3.99_{4.50}$	$_{3.25}\mathbf{3.92}_{4.50}$	$_{3.03}3.99_{4.50}$
6	$_{4.94}5.29_{5.56}$	$_{4.30}\mathbf{4.88}_{5.25}$	$_{4.28}4.89_{5.34}$	$_{4.30}\mathbf{4.88}_{5.25}$	$_{4.28}4.89_{5.34}$

$h_U^{(1)}, \ldots, h_U^{(5)}$, $U \in \{me, rme, ent, gini, R\}$, in the format $_{\min}Avg_{\max}$ (minimum values of Avg for each n are in bold).

The first column in Table 5.23 contains the number of variables n in the considered Boolean functions. The last five columns contain information about values $L_U^{(1)}, \ldots, L_U^{(5)}$, $U \in \{me, rme, ent, gini, R\}$, in the format $_{\min}Avg_{\max}$ (minimum values of Avg for each n are in bold).

Results of experiments for Boolean functions and decision rules are represented in Tables 5.24 and 5.25. The first column in Table 5.24 contains the number of variables n in the considered Boolean functions. The last five columns contain information about values $l_U^{(1)}, \ldots, l_U^{(5)}$, $U \in \{me, rme, ent, gini, R\}$, in the format $_{\min}Avg_{\max}$ (minimum values of Avg for each n are in bold).

The first column in Table 5.25 contains the number of variables n in the considered Boolean functions. The last five columns contain information about values $c_U^{(1)}, \ldots, c_U^{(5)}$,

Table 5.25 Experimental results for Boolean functions ($U \in \{me, rme, ent, gini, R\}$ and c)

Number of variables, n	$c_U^{(1)}$	$c_U^{(2)}$	$c_U^{(3)}$	$c_U^{(4)}$	$c_U^{(5)}$
3	${}_{1.00}1.85_{2.75}$	${}_{1.25}\mathbf{2.18}_{3.63}$	${}_{1.00}2.08_{3.63}$	${}_{1.25}\mathbf{2.18}_{3.63}$	${}_{1.00}2.08_{3.63}$
4	${}_{1.13}1.85_{4.63}$	${}_{1.31}\mathbf{2.54}_{6.44}$	${}_{1.50}2.26_{6.44}$	${}_{1.31}\mathbf{2.54}_{6.44}$	${}_{1.50}2.26_{6.44}$
5	${}_{1.38}1.86_{3.38}$	${}_{1.50}\mathbf{2.69}_{4.22}$	${}_{1.50}2.36_{4.84}$	${}_{1.50}\mathbf{2.69}_{4.22}$	${}_{1.50}2.36_{4.84}$
6	${}_{1.47}1.85_{2.75}$	${}_{2.00}\mathbf{2.86}_{4.17}$	${}_{1.67}2.60_{4.16}$	${}_{2.00}\mathbf{2.86}_{4.17}$	${}_{1.67}2.60_{4.16}$

$U \in \{me, rme, ent, gini, R\}$, in the format ${}_{\min}Avg_{\max}$ (maximum values of Avg for each n are in bold).

We now explain why the results of experiments on Boolean functions do not depend on the choice of uncertainty measures from the set $M = \{me, rme, ent, gini, R\}$.

Let $f(x_1, \ldots, x_n)$ be a nonconstant Boolean function, T_f be the decision table with attributes $x_1, \ldots, x_n$ representing f, and T be a nondegenerate subtable of T_f such that $T = T_f\{x_{j_1} = \delta_1, \ldots, x_{j_k} = \delta_k\}$, $k < n$, $1 \le j_1 < \cdots < j_k \le n$, and $\delta_1, \ldots, \delta_n \in \{0, 1\}$.

A subtable Θ of the table T_f is called 1-subtable of T if it can be represented in the form $\Theta = T\{x_i = \delta\}$, where $i \notin \{j_1, \ldots, j_k\}$ and $\delta \in \{0, 1\}$. This subtable contains exactly 2^t rows, where $t = n - k - 1$, i.e., $N(\Theta) = 2^t$. Denote $m(\Theta) = \min\{N_0(\Theta), N_1(\Theta)\}$. Then

- $me(\Theta) = m(\Theta)$.
- $rme(\Theta) = m(\Theta)/2^t$.
- $ent(\Theta) = -p \log_2 p - (1 - p) \log_2(1 - p)$, where $p = m(\Theta)/2^t$.
- $gini(\Theta) = m(\Theta)(2^t - m(\Theta))/2^{2t}$.
- $R(\Theta) = m(\Theta)(2^t - m(\Theta))/2$.

We will say that an uncertainty measure U is monotone for T if, for any 1-subtables Θ_1 and Θ_2 of T, $U(\Theta_1) \le U(\Theta_2)$ if and only if $m(\Theta_1) \le m(\Theta_2)$. We now show that each uncertainty measure from the set M is monotone for T.

For me and rme, the considered statement is obvious.

Let us consider the function $H(x) = -x \log_2 x - (1 - x) \log_2(1 - x)$, where x is a real number and $0 \le x \le 1$. It is well known that this function is increasing if $0 \le x \le 0.5$. Using this fact, it is easy to show that ent is a monotone uncertainty measure for T.

We now consider the function $r(x) = x(2^t - x)$, where x is an integer and $0 \le x \le 2^t$. One can show that this function is increasing if $0 \le x \le 2^{t-1}$. Using this fact, it is not difficult to show that $gini$ and R are monotone uncertainty measures for T.

Let $U \in M$ and let us assume that, for each pair Θ_1, Θ_2 of 1-subtables for T, we know if the inequality $U(\Theta_1) \le U(\Theta_2)$ holds or not. This information determines the sets of queries with the minimum impurity among (i) all admissible for T attributes, (ii) all admissible for T hypotheses, and (iii) all admissible for T attributes and hypotheses. Since all uncertainty measures from the set M are monotone for T, the sets of admissible for T queries with the minimum impurity do not depend on the choice of uncertainty measures from M.

We will not consider details of the software implementation for the greedy algorithms. However, the above reasoning allows us to understand independence of the experimental results for Boolean functions from the chosen uncertainty measure.

5.4 Analysis of Experimental Results

First, we evaluate results obtained for the decision tables from [8] based on average values of the parameters $h_U^{(k)}$, $L_U^{(k)}$, $l_U^{(k)}$, and $c_U^{(k)}$, where $U \in \{me, rme, ent, gini, R\}$ and $k = 1, \ldots, 5$.

To minimize the depth h of the constructed decision trees, we should choose decision trees of type 3. The two best uncertainty measures are R and me.

To minimize the number of realizable nodes L in the constructed decision trees, we should choose decision trees of type 1. The two best uncertainty measures are R and ent.

To minimize the length l of decision rules derived from the constructed decision trees, we should choose decision trees of type 2. The two best uncertainty measures are me and rme.

To maximize the coverage c of decision rules derived from the constructed decision trees, we should choose decision trees of type 4. The two best uncertainty measures are rme and me.

The results for randomly generated Boolean functions are consistent with the results for decision tables from [8]. For the minimization of h, we should use decision trees of types 3 and 5, for the minimization of L—trees of type 1, for the minimization of l—trees of types 2 and 4, and for the maximization of c—also trees of types 2 and 4. Note that the results obtained for Boolean functions do not depend on the choice of an uncertainty measure from the set $\{me, rme, ent, gini, R\}$.

5.5 Conclusions

In this chapter, we proposed greedy algorithms based on arbitrary uncertainty measures for the construction of decision trees with hypotheses and made experiments with five uncertainty measures. Using the results of experiments, we determined which types of decision trees and uncertainty measures should be chosen if we would like (i) to minimize the depth of decision trees, (ii) to minimize the number of realizable nodes in decision trees, (iii) to minimize the length of decision rules derived from the constructed decision trees, and (iv) to maximize the coverage of decision rules derived from the constructed decision trees.

References

1. AbouEisha, H., Amin, T., Chikalov, I., Hussain, S., Moshkov, M.: Extensions of Dynamic Programming for Combinatorial Optimization and Data Mining. Intelligent Systems Reference Library, vol. 146. Springer (2019)
2. Aglin, G., Nijssen, S., Schaus, P.: Learning optimal decision trees using caching branch-and-bound search. In: 34th AAAI Conference on Artificial Intelligence, AAAI 2020, pp. 3146–3153 (2020)
3. Alsolami, F., Azad, M., Chikalov, I., Moshkov, M.: Decision and Inhibitory Trees and Rules for Decision Tables with Many-Valued Decisions. Intelligent Systems Reference Library, vol. 156. Springer (2020)
4. Azad, M., Chikalov, I., Hussain, S., Moshkov, M.: Entropy-based greedy algorithm for decision trees using hypotheses. Entropy **23**(7), 808 (2021). https://doi.org/10.3390/e23070808
5. Azad, M., Chikalov, I., Hussain, S., Moshkov, M., Zielosko, B.: Decision rules derived from optimal decision trees with hypotheses. Entropy **23**(12), 1641 (2021). https://doi.org/10.3390/e23121641
6. Azad, M., Chikalov, I., Hussain, S., Moshkov, M., Zielosko, B.: Greedy algorithms for decision trees with hypotheses. arXiv:2203.08848 [cs.CC] (2022). https://doi.org/10.48550/arXiv.2203.08848
7. Breiman, L., Friedman, J.H., Olshen, R.A., Stone, C.J.: Classification and Regression Trees. Chapman and Hall/CRC (1984)
8. Dua, D., Graff, C.: UCI Machine Learning Repository. University of California, Irvine, School of Information and Computer Sciences (2017). http://archive.ics.uci.edu/ml
9. Narodytska, N., Ignatiev, A., Pereira, F., Marques-Silva, J.: Learning optimal decision trees with SAT. In: 27th International Joint Conference on Artificial Intelligence, IJCAI 2018, pp. 1362–1368 (2018)
10. Quinlan, J.R.: C4.5: Programs for Machine Learning. Morgan Kaufmann (1993)
11. Verwer, S., Zhang, Y.: Learning optimal classification trees using a binary linear program formulation. In: 33rd AAAI Conference on Artificial Intelligence, AAAI 2019, pp. 1625–1632 (2019)

6 Decision Trees with Hypotheses for Recognition of Monotone Boolean Functions and for Sorting

In this chapter, based on [2, 3], we look at using the tools described in Chaps. 3 and 5 to study the problem of recognition of monotone Boolean functions with n variables, $n = 2, \ldots, 4$, and the problem of sorting n pairwise different elements from linearly ordered set, $n = 3, \ldots, 6$. Note that these problems can be reformulated as exact learning problems.

For each of these problems, we compare the depth and the number of realizable nodes for five types of optimal (relative to the depth or the number of realizable nodes) decision trees. We compare the same parameters for five types of decision trees constructed by greedy algorithms based on the uncertainty measures me, rme, ent, $gini$, and R. We also analyze the length of decision rules derived from these trees.

For both problems, in almost each case we can find decision trees with hypotheses such that the results for these trees outperform the results for conventional decision trees. The only exception is the minimum number of realizable nodes for $n = 3, 4$ for the first problem (see Table 6.4) and for $n = 5, 6$ for the second problem (see Table 6.9).

For the problem of recognition of monotone Boolean functions, the results for the minimum depth of decision trees of type 1 (conventional decision trees) follow from well known result of Hansel [5]: the minimum depth of a conventional decision tree for the recognition of monotone Boolean functions with n variables is equal to $\binom{n}{\lfloor n/2 \rfloor} + \binom{n}{\lfloor n/2 \rfloor + 1}$. It is easy to show that the minimum number of nodes in a conventional decision tree for the recognition of monotone Boolean functions with n variables is equal to $2M(n) - 1$, where $M(n)$ is the number of monotone Boolean functions with n variables. Other results are new.

For the problem of sorting, the results for the optimal decision trees of type 1 are well known—see details in Sect. 6.1 of the book [1]. Other results are new.

The rest of the chapter is organized as follows. In Sect. 6.1, we consider the problem of recognition of monotone Boolean functions, and in Sect. 6.2—the problem of sorting. Section 6.3 contains short conclusions.

M. Azad et al., *Decision Trees with Hypotheses*, Synthesis Lectures on Intelligent Technologies, https://doi.org/10.1007/978-3-031-08585-7_6

6.1 Problem of Recognition of Monotone Boolean Functions

In this section, we study the problem of recognition of monotone Boolean functions with n variables, $n = 2, \ldots, 4$.

6.1.1 Basic Notions and Notation

A Boolean function $f(x_1, \ldots, x_n)$ is called *monotone* if, for any two tuples

$$(a_1, \ldots, a_n), (b_1, \ldots, b_n) \in \{0, 1\}^n$$

such that $a_1 \leq b_1, \ldots, a_n \leq b_n$, the inequality $f(a_1, \ldots, a_n) \leq f(b_1, \ldots, b_n)$ holds. For a given monotone Boolean function with n variables, we should recognize it using attributes each of which is the value of the considered function on some tuple from $\{0, 1\}^n$. We denote by $M(n)$ the number of monotone Boolean functions with n variables. It is known [4] that $M(2) = 6$, $M(3) = 20$, and $M(4) = 168$.

The problem of recognition of monotone Boolean functions with n variables $x_1, \ldots, x_n$ can be represented as a decision table R_n with 2^n conditional attributes $r_{a_1 \cdots a_n}$ corresponding to tuples $(a_1, \ldots, a_n)$ from $\{0, 1\}^n$ and $M(n)$ rows corresponding to monotone Boolean functions with n variables. For each such function f, the corresponding row of R_n is labeled with this function as the decision. In the intersection with the column labeled with the attribute $r_{a_1 \cdots a_n}$, the considered row has the value $f(a_1, \ldots, a_n)$.

The decision table R_2 is shown in Fig. 6.1. A decision tree of type 5 for the table R_2 is shown in Fig. 6.2 (we show only realizable nodes). The depth of this tree is equal to two and the number of realizable nodes is equal to nine.

The table R_n can also be considered as a representation of an exact learning problem: conditional attributes form a domain, each row describes a concept, and rows form a concept class.

We will use the following notation:

- $h^{(k)}(R_n)$ is the minimum depth of a decision tree of type k for R_n, $k = 1, \ldots, 5$.
- $L^{(k)}(R_n)$ is the minimum number of nodes realizable relative to R_n in a decision tree of type k for R_n, $k = 1, \ldots, 5$.

Fig. 6.1 Decision table R_2

$R_2 =$

r_{00}	r_{01}	r_{10}	r_{11}	
0	0	0	0	0
1	1	1	1	1
0	0	1	1	x_1
0	1	0	1	x_2
0	0	0	1	$x_1 \wedge x_2$
0	1	1	1	$x_1 \vee x_2$

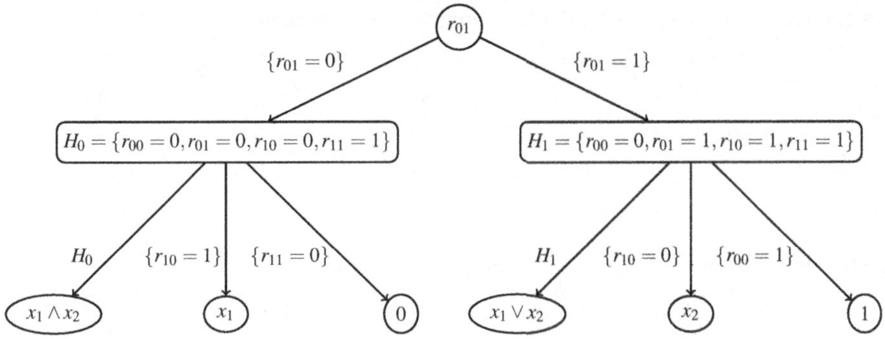

Fig. 6.2 Decision tree of type 5 for decision table R_2 (we show only realizable nodes)

Table 6.1 Experimental results for dynamic programming algorithm $\mathscr{A}_h$

n	$h^{(1)}(R_n)$	$h^{(2)}(R_n)$	$h^{(3)}(R_n)$	$h^{(4)}(R_n)$	$h^{(5)}(R_n)$
2	3	**2**	**2**	**2**	**2**
3	6	**3**	**3**	**3**	**3**
4	10	**6**	**6**	**6**	**6**

Let U be an uncertainty measure. The greedy algorithm $\mathscr{A}_U$ described in Chap. 5 constructs for the decision table R_n and $k \in \{1, \dots, 5\}$ a decision tree $\Gamma_U^{(k)}(R_n)$ of type k for the table R_n. We will use the following notation:

- $h_U^{(k)}(R_n) = h(\Gamma_U^{(k)}(R_n))$.
- $L_U^{(k)}(R_n) = L(R_n, \Gamma_U^{(k)}(R_n))$.
- $l_U^{(k)}(R_n) = l(R_n, \Gamma_U^{(k)}(R_n))$.

6.1.2 Results of Experiments

We make experiments with dynamic programming algorithms $\mathscr{A}_h$ and $\mathscr{A}_L$, and with greedy algorithms $\mathscr{A}_U$, $U \in \{me, rme, ent, gini, R\}$. Note that the results obtained for the greedy algorithms do not depend on the choice of the uncertainty measure U. We already met with similar situation in Chap. 5, when studied decision trees computing Boolean functions.

For $n = 2, \dots, 4$ and $k = 1, \dots, 5$, we find values of $h^{(k)}(R_n)$ and $L^{(k)}(R_n)$ using dynamic programming algorithms $\mathscr{A}_h$ and $\mathscr{A}_L$ described in Chap. 3—see results in Tables 6.1 and 6.2. Minimum values for each n are in bold.

The obtained experimental results show that the minimum depth of decision trees of types 2–5 is noticeably less than the minimum depth of decision trees of type 1. The minimum

Table 6.2 Experimental results for dynamic programming algorithm $\mathcal{A}_L$

n	$L^{(1)}(R_n)$	$L^{(2)}(R_n)$	$L^{(3)}(R_n)$	$L^{(4)}(R_n)$	$L^{(5)}(R_n)$
2	11	12	**9**	12	**9**
3	39	76	**33**	76	**33**
4	335	8808	**283**	8808	**283**

Table 6.3 Experimental results for greedy algorithm $\mathcal{A}_U$ (depth)

n	$h_U^{(1)}(R_n)$	$h_U^{(2)}(R_n)$	$h_U^{(3)}(R_n)$	$h_U^{(4)}(R_n)$	$h_U^{(5)}(R_n)$
2	3	**2**	**2**	**2**	**2**
3	6	**3**	**3**	**3**	**3**
4	10	**6**	**6**	**6**	**6**

Table 6.4 Experimental results for greedy algorithm $\mathcal{A}_U$ (number of realizable nodes)

n	$L_U^{(1)}(R_n)$	$L_U^{(2)}(R_n)$	$L_U^{(3)}(R_n)$	$L_U^{(4)}(R_n)$	$L_U^{(5)}(R_n)$
2	11	12	**9**	12	**9**
3	**39**	76	58	76	58
4	**335**	8850	1969	8850	1969

Table 6.5 Experimental results for greedy algorithm $\mathcal{A}_U$ (length of derived rules)

n	$l_U^{(1)}(R_n)$	$l_U^{(2)}(R_n)$	$l_U^{(3)}(R_n)$	$l_U^{(4)}(R_n)$	$l_U^{(5)}(R_n)$
2	2.67	**2.17**	2.33	**2.17**	2.33
3	4.55	3.50	**3.45**	3.50	**3.45**
4	7.65	**5.58**	5.94	**5.58**	5.94

number of realizable nodes in decision trees of types 3 and 5 is noticeably less than the minimum number of realizable nodes in decision trees of type 1. Decision trees of types 2 and 4 have too many nodes.

Let $U \in \{me, rme, ent, gini, R\}$. For $n = 2, \ldots, 4$ and $k = 1, \ldots, 5$, we construct a decision tree of type k for the decision table R_n by the greedy algorithm $\mathcal{A}_U$ described in Chap. 5 and find values $h_U^{(k)}(R_n)$, $L_U^{(k)}(R_n)$, and $l_U^{(k)}(R_n)$—see results in Tables 6.3, 6.4 and 6.5. Note that the obtained results do not depend on the choice of U. Minimum values for each n are in bold.

For the decision trees constructed by greedy algorithms, the obtained experimental results show that (i) for the depth, trees of types 2–5 outperform trees of type 1 for $n = 2, \ldots, 4$, (ii) for the number of realizable nodes, trees of types 3 and 5 outperform trees of type 1 for $n = 2$, and (iii) for the length of derived decision rules, trees of types 2–5 outperform trees of type 1 for $n = 2, \ldots, 4$.

6.2 Problem of Sorting

In this section, we study the problem of sorting n pairwise different elements from linearly ordered set, $n = 3, \ldots, 6$.

6.2.1 Basic Notions and Notation

Let $x_1, \ldots, x_n$ be pairwise different elements from a linearly ordered set. We should find a permutation $(p_1, \ldots, p_n)$ from the set P_n of all permutations of the set $\{1, \ldots, n\}$ for which $x_{p_1} < \cdots < x_{p_n}$. To this end, we use attributes $s_{i,j}$ such that $i, j \in \{1, \ldots, n\}, i < j$, $s_{i,j} = 1$ if $x_i < x_j$, and $s_{i,j} = 0$ if $x_i > x_j$.

The problem of sorting n elements can be represented as a decision table S_n with $n(n-1)/2$ conditional attributes $s_{i,j}$, $i, j \in \{1, \ldots, n\}$, $i < j$, and $n!$ rows corresponding to permutations from P_n. For each permutation $(p_1, \ldots, p_n)$, the corresponding row of S_n is labeled with this permutation as the decision. This row is filled with the values of the attributes $s_{i,j}$ such that $s_{i,j} = 1$ if and only if i comes before j in the tuple $(p_1, \ldots, p_n)$.

The decision table S_3 is shown in Fig. 6.3. A decision tree of type 5 for the table S_3 is shown in Fig. 6.4 (we show only realizable nodes). The depth of this tree is equal to two and the number of realizable nodes is equal to nine.

The table S_n can also be considered as a representation of an exact learning problem: conditional attributes form a domain, each row describes a concept, and rows form a concept class.

We will use the following notation:

- $h^{(k)}(S_n)$ is the minimum depth of a decision tree of type k for S_n, $k = 1, \ldots, 5$.
- $L^{(k)}(S_n)$ is the minimum number of nodes realizable relative to S_n in a decision tree of type k for S_n, $k = 1, \ldots, 5$.

Let U be an uncertainty measure. The greedy algorithm $\mathscr{A}_U$ described in Chap. 5 constructs for the decision table S_n and $k \in \{1, \ldots, 5\}$ a decision tree $\Gamma_U^{(k)}(S_n)$ of type k for the table S_n. We will use the following notation:

Fig. 6.3 Decision table S_3

$S_3 =$

$s_{1,2}$	$s_{1,3}$	$s_{2,3}$	
1	1	1	(1,2,3)
1	1	0	(1,3,2)
0	1	1	(2,1,3)
0	0	1	(2,3,1)
1	0	0	(3,1,2)
0	0	0	(3,2,1)

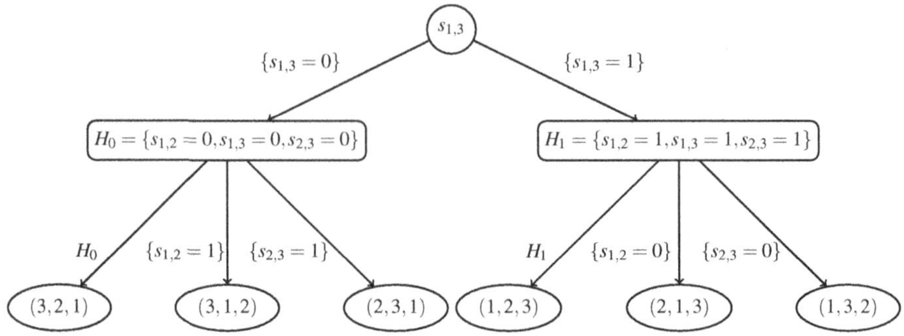

Fig. 6.4 Decision tree of type 5 for decision table S_3 (we show only realizable nodes)

- $h_U^{(k)}(S_n) = h(\Gamma_U^{(k)}(S_n))$.
- $L_U^{(k)}(S_n) = L(S_n, \Gamma_U^{(k)}(S_n))$.
- $l_U^{(k)}(S_n) = l(S_n, \Gamma_U^{(k)}(S_n))$.

6.2.2 Results of Experiments

We make experiments with dynamic programming algorithms $\mathscr{A}_h$ and $\mathscr{A}_L$, and with greedy algorithms $\mathscr{A}_U$, $U \in \{me, rme, ent, gini, R\}$. Note that the results obtained for the greedy algorithms do not depend on the choice of the uncertainty measure U. We already met with similar situation in Chap. 5, when studied decision trees computing Boolean functions, and in the previous section, when studied decision trees for the recognition of monotone Boolean functions.

For $n = 3, \ldots, 6$ and $k = 1, \ldots, 5$, we find values of $h^{(k)}(S_n)$ and $L^{(k)}(S_n)$ using dynamic programming algorithms $\mathscr{A}_h$ and $\mathscr{A}_L$ described in Chap. 3—see results in Tables 6.6 and 6.7. Minimum values for each n are in bold.

From the obtained experimental results it follows that the minimum depth of decision trees of types 2–5 is one less than the minimum depth of decision trees of type 1. The minimum number of realizable nodes in decision trees of types 3 and 5 is noticeably less

Table 6.6 Experimental results for dynamic programming algorithm $\mathscr{A}_h$

n	$h^{(1)}(S_n)$	$h^{(2)}(S_n)$	$h^{(3)}(S_n)$	$h^{(4)}(S_n)$	$h^{(5)}(S_n)$
3	3	**2**	**2**	**2**	**2**
4	5	**4**	**4**	**4**	**4**
5	7	**6**	**6**	**6**	**6**
6	10	**9**	**9**	**9**	**9**

Table 6.7 Experimental results for dynamic programming algorithm $\mathcal{A}_L$

n	$L^{(1)}(S_n)$	$L^{(2)}(S_n)$	$L^{(3)}(S_n)$	$L^{(4)}(S_n)$	$L^{(5)}(S_n)$
3	11	13	**9**	14	**9**
4	47	253	**39**	254	**39**
5	239	15,071	**199**	15,142	**199**
6	1439	2,885,086	**1199**	2,886,752	**1199**

Table 6.8 Experimental results for greedy algorithm $\mathcal{A}_U$ (depth)

n	$h_U^{(1)}(S_n)$	$h_U^{(2)}(S_n)$	$h_U^{(3)}(S_n)$	$h_U^{(4)}(S_n)$	$h_U^{(5)}(S_n)$
3	3	**2**	**2**	**2**	**2**
4	5	**4**	**4**	**4**	**4**
5	7	**6**	**6**	**6**	**6**
6	10	**9**	**9**	**9**	**9**

Table 6.9 Experimental results for greedy algorithm $\mathcal{A}_U$ (number of realizable nodes)

n	$L_U^{(1)}(S_n)$	$L_U^{(2)}(S_n)$	$L_U^{(3)}(S_n)$	$L_U^{(4)}(S_n)$	$L_U^{(5)}(S_n)$
3	11	14	**9**	14	**9**
4	47	254	**39**	254	**39**
5	**239**	15,142	455	15,142	455
6	**1439**	2,898,512	7231	2,898,512	7231

Table 6.10 Experimental results for greedy algorithm $\mathcal{A}_U$ (length of derived rules)

n	$l_U^{(1)}(S_n)$	$l_U^{(2)}(S_n)$	$l_U^{(3)}(S_n)$	$l_U^{(4)}(S_n)$	$l_U^{(5)}(S_n)$
3	2.67	**2.17**	2.33	**2.17**	2.33
4	4.67	**3.13**	4.33	**3.13**	4.33
5	6.93	**4.05**	6.13	**4.05**	6.13
6	9.58	**5.01**	7.96	**5.01**	7.96

than the minimum number of realizable nodes in decision trees of type 1. Decision trees of types 2 and 4 have too many nodes.

Let $U \in \{me, rme, ent, gini, R\}$. For $n = 3, \ldots, 6$ and $k = 1, \ldots, 5$, we construct a decision tree of type k for the decision table S_n by the greedy algorithm $\mathcal{A}_U$ described in Chap. 5 and find values $h_U^{(k)}(S_n)$, $L_U^{(k)}(S_n)$, and $l_U^{(k)}(S_n)$—see results in Tables 6.8, 6.9 and 6.10. Note that the obtained results do not depend on the choice of U. Minimum values for each n are in bold.

For the decision trees constructed by greedy algorithms, the obtained experimental results show that (i) for the depth, trees of types 2–5 outperform trees of type 1 for $n = 3, \ldots, 6$, (ii) for the number of realizable nodes, trees of types 3 and 5 outperform trees of type 1 for $n = 3, 4$, and (iii) for the length of derived decision rules, trees of types 2–5 (especially trees of types 2 and 4) outperform trees of type 1 for $n = 3, \ldots, 6$.

6.3 Conclusions

In this chapter, we studied two problems: the problem of recognition of monotone Boolean functions with n variables, $n = 2, \ldots, 4$, and the problem of sorting n pairwise different elements from linearly ordered set, $n = 3, \ldots, 6$. For each problem, we found the minimum depth and the minimum number of realizable nodes in decision trees solving the problem. We also found the depth and the number of realizable nodes in decision trees constructed by greedy algorithms based on the uncertainty measures me, rme, ent, $gini$, and R, and study the length of decision rules derived from these trees.

References

1. AbouEisha, H., Amin, T., Chikalov, I., Hussain, S., Moshkov, M.: Extensions of Dynamic Programming for Combinatorial Optimization and Data Mining. Intelligent Systems Reference Library, vol. 146. Springer (2019)
2. Azad, M., Chikalov, I., Hussain, S., Moshkov, M.: Sorting by decision trees with hypotheses (extended abstract). In: Schlingloff, H., Vogel, T. (eds.) 29th International Workshop on Concurrency, Specification and Programming, CS&P 2021, Berlin, Germany, 27–28 Sept 2021. CEUR Workshop Proceedings, vol. 2951, pp. 126–130. CEUR-WS.org (2021). http://ceur-ws.org/Vol-2951/paper1.pdf
3. Azad, M., Chikalov, I., Hussain, S., Moshkov, M., Zielosko, B.: Decision trees with hypotheses for recognition of monotone Boolean functions and for sorting. arXiv:2203.08894 [cs.CC] (2022). https://doi.org/10.48550/arXiv.2203.08894
4. Church, R.: Numerical analysis of certain free distributive structures. Duke Math. J. **6**(3), 732–734 (1940)
5. Hansel, G.: Sur le nombre des fonctions booléennes monotones de n variables. C. R. Acad. Sci. Paris **262**(20), 1088–1090 (1966)

Binary Information Systems and Infinite Families of Concepts

Part II

In this part, we study arbitrary infinite binary information systems each of which consists of an infinite set of elements and an infinite set of two-valued functions (attributes) defined on the set of elements. We consider the notion of a problem over information system, which is described by a finite number of attributes: for a given element, we should recognize values of these attributes. For an arbitrary infinite binary information system, we study five functions of Shannon type, which characterize the dependence in the worst case of the minimum depth of a decision tree solving a problem on the number of attributes in the problem description. These five functions correspond to (1) decision trees using attributes, (2) decision trees using hypotheses, (3) decision trees using both attributes and hypotheses, (4) decision trees using proper hypotheses, and (5) decision trees using both attributes and proper hypotheses. For each of these functions, we describe possible types of behavior. We also study joint behavior of the considered five functions and distinguish seven complexity classes of infinite binary information systems.

We translate the obtained results into exact learning. To this end, we study arbitrary infinite families of concepts each of which consists of an infinite set of elements and an infinite set of subsets of this set called concepts. We consider the notion of a problem over a family of concepts that is described by a finite number of elements: for a given concept, we should recognize which of the elements under consideration belong to this concept. As algorithms for problem solving, we study decision trees of five types: (1) using membership queries, (2) using equivalence queries, (3) using both membership and equivalence queries, (4) using proper equivalence queries, and (5) using both membership and proper equivalence queries.

In Chap. 7, we investigate decision trees of types 1, 2, and 3 over infinite binary information systems.

In Chap. 8, we study decision trees of types 4 and 5 over infinite binary information systems.

In Chap. 9, we translate results obtained in Chaps. 7 and 8 into exact learning and study infinite families of concepts.

Infinite Binary Information Systems. Decision Trees of Types 1, 2, and 3

7

In this chapter, we begin our study of decision trees over infinite binary information systems. This chapter is based on the results published in [7–10].

Each infinite binary information systems consists of an infinite set of elements A and an infinite set F of functions (attributes) from A to $\{0, 1\}$. We define the notion of a problem described by a finite number of attributes $f_1, \ldots, f_n$ from F: for a given element $a \in A$, we should recognize the tuple $(f_1(a), \ldots, f_n(a))$. To this end, we can use decision trees based on two types of queries. We can ask about the value of an attribute $f_i \in \{f_1, \ldots, f_n\}$. We will obtain an answer of the kind $f_i(x) = \delta$ where $\delta \in \{0, 1\}$. We can also ask if a hypothesis $f_1(x) = \delta_1, \ldots, f_n(x) = \delta_n$ is true where $\delta_1, \ldots, \delta_n \in \{0, 1\}$. Either this hypothesis will be confirmed or we will obtain a counterexample in the form $f_i(x) = \neg\delta_i$. As time complexity of a decision tree, we consider its depth, which is equal to the maximum number of queries in a path from the root to a terminal node of the tree.

Based on the results of exact learning [1–4] and test theory and rough set theory [6, 7, 9], for an arbitrary infinite binary information system, we study three functions of Shannon type, which characterize the dependence in the worst case of the minimum depth of a decision tree solving a problem on the number of attributes in the problem description. The considered three functions correspond to the following three types of decision trees:

(1) Only attributes are used in decision trees.
(2) Only hypotheses are used in decision trees.
(3) Both attributes and hypotheses are used in decision trees.

We show that the first function has two possible types of behavior: logarithmic and linear. The second and third functions have three possible types of behavior: constant, logarithmic, and linear. We also investigate join behavior of the considered three Shannon functions and describe four complexity classes of infinite binary information systems.

M. Azad et al., *Decision Trees with Hypotheses*, Synthesis Lectures on Intelligent Technologies, https://doi.org/10.1007/978-3-031-08585-7_7

Bounds for the decision trees of type 1 can be derived from the results obtained in [7, 9]. In the present chapter, we provide a direct proof for these bounds. Results for the decision trees of types 2 and 3 were presented in the conference paper [8] without proofs. Proofs for these results and results related to the join behavior of the three Shannon type functions were published in [10].

The rest of the chapter is organized as follows. In Sects. 7.1 and 7.2, we consider basic notions and main results. Sections 7.3 and 7.4 contains proofs of main results, and Sect. 7.5—short conclusions.

7.1 Basic Notions

Let A be a set and F be a set of functions from A to $\{0, 1\}$. Functions from F are called *attributes* and the pair $U = (A, F)$ is called a *binary information system* (this notion is close to the notion of information system proposed by Pawlak [12]). If A and F are infinite sets, then the pair $U = (A, F)$ is called an *infinite binary information system.*

The set A may be interpreted as the set of inputs for problems over the information system U. A *problem over* U is an arbitrary n-tuple $z = (f_1, \dots, f_n)$ where $n \in \mathbb{N}$, $\mathbb{N}$ is the set of natural numbers $\{1, 2, \dots\}$, and $f_1, \dots, f_n \in F$. The problem z may be interpreted as a problem of searching for the tuple $z(a) = (f_1(a), \dots, f_n(a))$ for an arbitrary $a \in A$. The number $\dim z = n$ is called the *dimension* of the problem z. Denote $F(z) = \{f_1, \dots, f_n\}$. We denote by $P(U)$ the set of problems over U.

A *system of equations over* U is an arbitrary equation system of the kind

$$\{g_1(x) = \delta_1, \dots, g_m(x) = \delta_m\}$$

where $m \in \mathbb{N} \cup \{0\}$, $g_1, \dots, g_m \in F$, and $\delta_1, \dots, \delta_m \in \{0, 1\}$ (if $m = 0$, then the considered equation system is empty). This equation system is called a *system of equations over* z if $g_1, \dots, g_m \in F(z)$. The considered equation system is called *consistent* (on A) if its set of solutions on A is nonempty. The set of solutions of the empty equation system coincides with A.

As algorithms for problem z solving, we consider decision trees with two types of *queries*. We can choose an attribute $f_i \in F(z)$ and ask about its value. This query has two possible answers $\{f_i(x) = 0\}$ and $\{f_i(x) = 1\}$. We can formulate a *hypothesis over* z in the form $H = \{f_1(x) = \delta_1, \dots, f_n(x) = \delta_n\}$ where $\delta_1, \dots, \delta_n \in \{0, 1\}$, and ask about this hypothesis. This query has $n + 1$ possible answers: $H, \{f_1(x) = \neg\delta_1\}, \dots, \{f_n(x) = \neg\delta_n\}$ where $\neg 1 = 0$ and $\neg 0 = 1$. The first answer means that the hypothesis is true. Other answers are counterexamples. The hypothesis H is called *proper* if the system of equations H is consistent. In this chapter, we consider arbitrary hypotheses. Proper hypotheses will be considered in the next chapter.

A *decision tree over* z is a marked finite directed tree with the root in which

- Each *terminal* node is labeled with an n-tuple from the set $\{0, 1\}^n$.
- Each node, which is not terminal (such nodes are called *working*), is labeled with an attribute from the set $F(z)$ or with a hypothesis over z.
- If a working node is labeled with an attribute f_i from $F(z)$, then there are two edges, which leave this node and are labeled with the systems of equations $\{f_i(x) = 0\}$ and $\{f_i(x) = 1\}$, respectively.
- If a working node is labeled with a hypothesis

$$H = \{f_1(x) = \delta_1, \dots, f_n(x) = \delta_n\}$$

over z, then there are $n + 1$ edges, which leave this node and are labeled with the systems of equations H, $\{f_1(x) = \neg\delta_1\}$, ..., $\{f_n(x) = \neg\delta_n\}$, respectively.

Let Γ be a decision tree over z. A *complete path* in Γ is an arbitrary directed path from the root to a terminal node in Γ. We now define an equation system $\mathcal{S}(\xi)$ over U associated with the complete path ξ. If there are no working nodes in ξ, then $\mathcal{S}(\xi)$ is the empty system. Otherwise, $\mathcal{S}(\xi)$ is the union of equation systems assigned to the edges of the path ξ. We denote by $\mathcal{A}(\xi)$ the set of solutions on A of the system of equations $\mathcal{S}(\xi)$ (if this system is empty, then its solution set is equal to A).

We will say that a decision tree Γ over z *solves the problem* z *relative to* U if, for each element $a \in A$ and for each complete path ξ in Γ such that $a \in \mathcal{A}(\xi)$, the terminal node of the path ξ is labeled with the tuple $z(a)$.

We now consider an equivalent definition of a decision tree solving a problem. Denote by $\Delta_U(z)$ the set of tuples $(\delta_1, \dots, \delta_n) \in \{0, 1\}^n$ such that the system of equations $\{f_1(x) = \delta_1, \dots, f_n(x) = \delta_n\}$ is consistent. The set $\Delta_U(z)$ is the set of all possible solutions to the problem z. Let $\Delta \subseteq \Delta_U(z)$, $f_{i_1}, \dots, f_{i_m} \in \{f_1, \dots, f_n\}$, and $\sigma_1, \dots, \sigma_m \in \{0, 1\}$. Denote

$$\Delta(f_{i_1}, \sigma_1) \cdots (f_{i_m}, \sigma_m)$$

the set of all n-tuples $(\delta_1, \dots, \delta_n) \in \Delta$ for which $\delta_{i_1} = \sigma_1, \dots, \delta_{i_m} = \sigma_m$.

Let Γ be a decision tree over the problem z. We correspond to each complete path ξ in the tree Γ a word $\pi(\xi)$ in the alphabet $\{(f_i, \delta) : f_i \in F(z), \delta \in \{0, 1\}\}$. If the equation system $\mathcal{S}(\xi)$ is empty, then $\pi(\xi)$ is the empty word. If $\mathcal{S}(\xi) = \{f_{i_1}(x) = \sigma_1, \dots, f_{i_m}(x) = \sigma_m\}$, then $\pi(\xi) = (f_{i_1}, \sigma_1) \cdots (f_{i_m}, \sigma_m)$. The decision tree Γ over z solves the problem z relative to U if, for each complete path ξ in Γ, the set $\Delta_U(z)\pi(\xi)$ contains at most one tuple and if this set contains exactly one tuple, then the considered tuple is assigned to the terminal node of the path ξ.

As time complexity of a decision tree, we consider its *depth* that is the maximum number of working nodes in a complete path in the tree or, which is the same, the maximum length of a complete path in the tree. We denote by $h(\Gamma)$ the depth of a decision tree Γ.

Let $z \in P(U)$. We denote by $h_U^{(1)}(z)$ the minimum depth of a decision tree over z, which solves z relative to U and uses only attributes from $F(z)$. We denote by $h_U^{(2)}(z)$ the minimum depth of a decision tree over z, which solves z relative to U and uses only hypotheses over z. We denote by $h_U^{(3)}(z)$ the minimum depth of a decision tree over z, which solves z relative to U and uses both attributes from $F(z)$ and hypotheses over z.

For $i = 1, 2, 3$, we define a function of Shannon type $h_U^{(i)}(n)$ that characterizes dependence of $h_U^{(i)}(z)$ on $\dim z$ in the worst case. Let $i \in \{1, 2, 3\}$ and $n \in \mathbb{N}$. Then

$$h_U^{(i)}(n) = \max\{h_U^{(i)}(z) : z \in P(U), \dim z \leq n\} .$$

7.2 Main Results

Let $U = (A, F)$ be an infinite binary information system and $r \in \mathbb{N}$. We will say that the information system U is *r-reduced* if, for each consistent on A system of equations over U, there exists a subsystem of this system that has the same set of solutions and contains at most r equations. We denote by $\mathscr{R}$ the class of infinite binary information systems each of which is *r-reduced* for some $r \in \mathbb{N}$.

The next theorem follows from results obtained in [7] where we considered closed classes of test tables (decision tables). It also follows from the results obtained in [9] where we considered the weighted depth of decision trees. For the completeness, we will give the proof of this theorem in Sect. 7.3.

Theorem 7.1 *Let U be an infinite binary information system. Then the following statements hold:*

(a) If $U \in \mathscr{R}$, then $h_U^{(1)}(n) = \Theta(\log n)$.

(b) If $U \notin \mathscr{R}$, then $h_U^{(1)}(n) = n$ for any $n \in \mathbb{N}$.

A subset $\{f_1, \ldots, f_m\}$ of F is called *independent* if, for any $\delta_1, \ldots, \delta_m \in \{0, 1\}$, the system of equations $\{f_1(x) = \delta_1, \ldots, f_m(x) = \delta_m\}$ is consistent on the set A. The empty set of attributes is independent by definition. We now define the parameter $I(U)$, which is called the *independence dimension* or *I-dimension* of the information system U (this notion is similar to the notion of independence number of family of sets considered by Naiman and Wynn in [11]). If, for each $m \in \mathbb{N}$, the set F contains an independent subset of the cardinality m, then $I(U) = \infty$. Otherwise, $I(U)$ is the maximum cardinality of an independent subset of the set F. We denote by $\mathscr{D}$ the class of infinite binary information systems with finite independence dimension.

Let $U = (A, F)$ be a binary information system, which is not necessary infinite, $f \in F$, and $\delta \in \{0, 1\}$. Denote

$$A(f, \delta) = \{a : a \in A, f(a) = \delta\} .$$

We now define inductively the notion of *k-information system*, $k \in \mathbb{N} \cup \{0\}$. The binary information system U is called 0-information system if all attributes from F are constant

Table 7.1 Possible indicator vectors of infinite binary information systems

	$\mathscr{R}$	$\mathscr{D}$	$\mathscr{C}$
1	0	0	0
2	0	1	0
3	0	1	1
4	1	1	0

on the set A. Let, for some $k \in \mathbb{N} \cup \{0\}$, the notion of m-information system be defined for $m = 0, \ldots, k$. The binary information system U is called $(k+1)$-information system if it is not m-information system for $m = 0, \ldots, k$ and, for any $f \in F$, there exist numbers $\delta \in \{0, 1\}$ and $m \in \{0, \ldots, k\}$ such that the information system $(A(f, \delta), F)$ is m-information system. It is easy to show by induction on k that if $U = (A, F)$ is k-information system, then $U' = (A', F)$, $A' \subseteq A$, is l-information system for some $l \leq k$. We denote by $\mathscr{C}$ the class of infinite binary information systems for each of which there exists $k \in \mathbb{N}$ such that the considered system is k-information system. Later we will show (see Proposition 7.2) that $\mathscr{C} \subseteq \mathscr{D}$. The following theorem was presented in [8] without proof.

Theorem 7.2 *Let U be an infinite binary information system. Then the following statements hold:*

(a) If $U \in \mathscr{C}$, then $h_U^{(2)}(n) = O(1)$ and $h_U^{(3)}(n) = O(1)$.

(b) If $U \in \mathscr{D} \setminus \mathscr{C}$, then $h_U^{(2)}(n) = \Theta(\log n)$, $h_U^{(3)}(n) = \Omega(\frac{\log n}{\log \log n})$, and $h_U^{(3)}(n) = O(\log n)$.

(c) If $U \notin \mathscr{D}$, then $h_U^{(2)}(n) = n$ and $h_U^{(3)}(n) = n$ for any $n \in \mathbb{N}$.

Let U be an infinite binary information system. We now consider the join behavior of the functions $h_U^{(1)}(n)$, $h_U^{(2)}(n)$, and $h_U^{(3)}(n)$. It depends on the belonging of the information system U to the classes $\mathscr{R}$, $\mathscr{D}$, and $\mathscr{C}$. We correspond to the information system U its *indicator vector* $ind(U) = (c_1, c_2, c_3) \in \{0, 1\}^3$ in which $c_1 = 1$ if and only if $U \in \mathscr{R}$, $c_2 = 1$ if and only if $U \in \mathscr{D}$, and $c_3 = 1$ if and only if $U \in \mathscr{C}$.

Theorem 7.3 *For any infinite binary information system, its indicator vector coincides with one of the rows of Table 7.1. Each row of Table 7.1 is the indicator vector of some infinite binary information system.*

For $i = 1, 2, 3, 4$, we denote by V_i the class of all infinite binary information systems, which indicator vector coincides with the ith row of Table 7.1. Table 7.2 summarizes Theorems 7.1–7.3. The first column contains the name of *complexity class* V_i. The next three columns describe the indicator vector of information systems from this class. The last three

Table 7.2 Summary of Theorems 7.1–7.3

	$\mathscr{R}$	$\mathscr{D}$	$\mathscr{C}$	$h_U^{(1)}(n)$	$h_U^{(2)}(n)$	$h_U^{(3)}(n)$
V_1	0	0	0	n	n	n
V_2	0	1	0	n	$\Theta(\log n)$	$\Omega(\frac{\log n}{\log\log n})$, $O(\log n)$
V_3	0	1	1	n	$O(1)$	$O(1)$
V_4	1	1	0	$\Theta(\log n)$	$\Theta(\log n)$	$\Omega(\frac{\log n}{\log\log n})$, $O(\log n)$

columns $h_U^{(1)}(n)$, $h_U^{(2)}(n)$, and $h_U^{(3)}(n)$ contain information about behavior of the functions $h_U^{(1)}(n)$, $h_U^{(2)}(n)$, and $h_U^{(3)}(n)$ for information systems from the class V_i.

7.3 Proofs of Theorems 7.1 and 7.2

In this section, we prove Theorems 7.1 and 7.2.

Proof (of Theorem 7.1) (a) Let $U \in \mathscr{R}$. First, we prove that $h_U^{(1)}(n) = O(\log n)$. Since $U \in \mathscr{R}$, there exists a natural r such that, for each consistent on A system of equations over U, there exists a subsystem of this system, which has the same set of solutions on A and contains at most r equations.

Let $z = (f_1, \dots, f_m)$ be a problem over U. The number of equation systems over z containing at most r equations is at most the number of r-tuples of equations of the kind $f_i = \delta$ where $f_i \in \{f_1, \dots, f_m\}$ and $\delta \in \{0, 1\}$. The latter number is equal to $(2m)^r$. Therefore $|\Delta_U(z)| \le (2m)^r$.

We consider a decision tree Γ, which solves the problem z relative to U and uses only attributes. This tree is constructed by a halving algorithm that is similar to proposed in [5, 6]. We will describe the work of Γ for an arbitrary element a from A. This work consists of the following steps.

Denote $\Delta = \Delta_U(z)$. If $|\Delta| = 1$, then the only n-tuple from Δ is the solution $z(a)$ of the problem z for the element a. Let $|\Delta| \ge 2$. For $i = 1, \dots, m$, we denote by δ_i a number from $\{0, 1\}$ such that $|\Delta(f_i, \delta_i)| \ge |\Delta(f_i, \neg\delta_i)|$. Let k be the maximum number from $\{1, \dots, m\}$ for which the equation system $\{f_1(x) = \delta_1, \dots, f_k(x) = \delta_k\}$ is consistent on A. This system contains a subsystem $\{f_{i_1}(x) = \delta_{i_1}, \dots, f_{i_t}(x) = \delta_{i_t}\}$, which has the same set of solutions on A and for which $t \le r$. We now show that $|\Delta(f_{i_1}, \delta_{i_1}) \cdots (f_{i_t}, \delta_{i_t})| \le |\Delta|/2$. If $k = m$, then $|\Delta(f_{i_1}, \delta_{i_1}) \cdots (f_{i_t}, \delta_{i_t})| \le 1 \le |\Delta|/2$. Let $k < m$. Then the equation system $\{f_{i_1}(x) = \delta_{i_1}, \dots, f_{i_t}(x) = \delta_{i_t}, f_{k+1}(x) = \delta_{k+1}\}$ is inconsistent. Therefore $\Delta(f_{i_1}, \delta_{i_1}) \cdots (f_{i_t}, \delta_{i_t}) \subseteq \Delta(f_{k+1}, \neg\delta_{k+1})$ and

$$|\Delta(f_{i_1}, \delta_{i_1}) \cdots (f_{i_t}, \delta_{i_t})| \le |\Delta|/2 .$$

We sequentially compute values of the attributes $f_{i_1}, \ldots, f_{i_t}$ for the element a. If $f_{i_1}(a) = \delta_{i_1}, \ldots, f_{i_t}(a) = \delta_{i_t}$, then during the next step we will work with the set of tuples $\Delta' = \Delta(f_{i_1}, \delta_{i_1}) \cdots (f_{i_t}, \delta_{i_t})$ for which $|\Delta'| \leq |\Delta|/2$. If, for some $p \in \{1, \ldots, t-1\}$, $f_{i_1}(a) = \delta_{i_1}, \ldots, f_{i_p}(a) = \delta_{i_p}$ and $f_{i_{p+1}}(a) = \neg\delta_{i_{p+1}}$, then during the next step we will work with the set of tuples

$$\Delta'' = \Delta(f_{i_1}, \delta_{i_1}) \cdots (f_{i_p}, \delta_{i_p})(f_{i_{p+1}}, \neg\delta_{i_{p+1}}) .$$

It is clear that $|\Delta''| \leq |\Delta|/2$. At this step, we compute values of at most r attributes and reduce the number of m-tuples (possible solutions) by half.

Let during the work with the element a, the decision tree Γ take q steps. After $(q-1)$th step, the number of remaining m-tuples will be at least two and at most $(2m)^r/2^{q-1}$. Therefore $2^q \leq (2m)^r$ and $q \leq r \log_2(2m)$. So during the processing of the element a, the decision tree Γ computes values of at most $r^2 \log_2(2m)$ attributes. Since a is an arbitrary element from A, the depth of Γ is at most $r^2 \log_2(2m)$. Since z is an arbitrary problem over U, we obtain $h_U^{(1)}(n) = O(\log n)$.

We now show that $h_U^{(1)}(n) = \Omega(\log n)$. We prove by induction on n that, for any natural n, there is a problem $z_n = (f_1, \ldots, f_n)$ over U such that $|\Delta_U(z_n)| \geq n+1$. Since F is an infinite set of attributes, there exists an attribute $f_1 \in F$ such that f_1 is not constant on the set A. Therefore $|\Delta_U(z_1)| \geq 2$ where $z_1 = (f_1)$. Let, for some natural n, there exist a problem $z_n = (f_1, \ldots, f_n)$ over U such that $|\Delta_U(z_n)| \geq n+1$. We now show that there exists an attribute $f_{n+1} \in F$ for which $|\Delta_U(z_{n+1})| \geq n+2$ where $z_{n+1} = (f_1, \ldots, f_n, f_{n+1})$. Assume the contrary: for any attribute f from F, we have $|\Delta_U(z')| = n+1$ where $z' = (f_1, \ldots, f_n, f)$. In this case, the set of attributes F is finite, which is impossible. Let $n \in \mathbb{N}$, $z_n = (f_1, \ldots, f_n)$ be a problem over U such that $|\Delta_U(z_n)| \geq n+1$, and Γ be a decision tree solving the problem z relative to U. Then Γ should have at least $n+1$ terminal nodes. One can show that the number of terminal nodes in the tree Γ is at most $2^{h(\Gamma)}$. Therefore $n+1 \leq 2^{h(\Gamma)}$, $h(\Gamma) \geq \log_2(n+1)$, and $h_U(z) \geq \log_2(n+1)$. Thus, $h_U^{(1)}(n) = \Omega(\log n)$ and $h_U^{(1)}(n) = \Theta(\log n)$.

(b) Let $U \notin \mathscr{R}$. One can show that, for any $n \in \mathbb{N}$, there is a consistent system of equations $S = \{f_1(x) = \delta_1, \ldots, f_n(x) = \delta_n\}$ over U for which each proper subsystem has different solution set than S. Let Γ be a decision tree solving the problem $z = (f_1, \ldots, f_n)$ relative to U. Then there is a complete path ξ in Γ such that $\Delta_U(z_n)\pi(\xi) = \{(\delta_1, \ldots, \delta_n)\}$. From here it follows that $S = \mathscr{S}(\xi)$. Therefore the complete path ξ contains at least n working nodes and $h(\Gamma) \geq n$. Taking into account that Γ is an arbitrary decision tree solving z relative to U, we obtain $h_U^{(1)}(z) \geq n$ and $h_U^{(1)}(n) \geq n$. It is easy to show that, for each problem z over U, the inequality $h_U^{(1)}(z) \leq \dim z$ holds: to solve the problem z, it is enough to compute values of all attributes from the set $F(z)$. Thus, $h_U^{(1)}(n) = n$ for any $n \in \mathbb{N}$. □

We precede the proof of Theorem 7.2 by two lemmas.

Let $d \in \mathbb{N}$. A *d-complete tree over the information system $U = (A, F)$* is a marked finite directed tree with the root in which

- Each terminal node is not labeled.
- Each nonterminal node is labeled with an attribute $f \in F$. There are two edges leaving this node that are labeled with the systems of equations $\{f(x) = 0\}$ and $\{f(x) = 1\}$, respectively.
- The length of each complete path (path from the root to a terminal node) is equal to d.
- For each complete path ξ, the equation system $\mathscr{S}(\xi)$, which is the union of equation systems assigned to the edges of the path ξ, is consistent.

Let G be a d-complete tree over U and $F(G)$ be the set of all attributes attached to the nonterminal nodes of the tree G. The number of nonterminal nodes in G is equal to $2^0 + 2^1 + \ldots + 2^{d-1} = 2^d - 1$. Therefore $|F(G)| \leq 2^d$.

The results mentioned in the following lemma are obtained by methods similar to used by Littlestone [3], Maass and Turán [4], and Angluin [2].

Lemma 7.1 *Let $U = (A, F)$ be a binary information system, $d \in \mathbb{N}$, G be a d-complete tree over U, and z be a problem over U such that $F(G) \subseteq F(z)$. Then*
(a) $h_U^{(2)}(z) \geq d$.
(b) $h_U^{(3)}(z) \geq \frac{d}{\log_2(2d)}$.

Proof (a) We prove the inequality $h_U^{(2)}(z) \geq d$ by induction on d. Let $d = 1$. Then the tree G has the only one nonterminal node, which is labeled with an attribute f that is not constant on A. Therefore $|\Delta_U(z)| \geq 2$ and $h_U^{(2)}(z) \geq 1$. Let, for $t \in \mathbb{N}$ and for any natural d, $1 \leq d \leq t$, the considered statement hold. Assume now that $d = t + 1$, G is a d-complete tree over U, z is a problem over U such that $F(G) \subseteq F(z)$, and Γ is a decision tree over z with the minimum depth, which solves the problem z and uses only hypotheses. Let f be the attribute attached to the root of the tree G and H be the hypothesis attached to the root of the decision tree Γ. Then there is an edge, which leaves the root of Γ and is labeled with the equation system $\{f(x) = \delta\}$ where the equation $f(x) = \neg\delta$ belongs to the hypothesis H. This edge enters to the root of the subtree of Γ, which will be denoted by Γ_f. There is an edge, which leaves the root of G and is labeled with the equation system $\{f(x) = \delta\}$. This edge enters to the root of the subtree of G, which will be denoted by G_δ. One can show that the decision tree Γ_f solves the problem z relative to the information system $U' = (A(f, \delta), F)$ and G_δ is a t-complete tree over U'. It is clear that $F(G_\delta) \subseteq F(z)$. Using the inductive hypothesis, we obtain $h(\Gamma_f) \geq t$. Therefore $h(\Gamma) \geq t + 1 = d$ and $h_U^{(2)}(z) \geq d$.

(b) We now prove the inequality $h_U^{(3)}(z) \geq \frac{d}{\log_2(2d)}$. Let $z = (f_1, \ldots, f_n)$ and Γ be a decision tree over z with the minimum depth, which solves the problem z and uses both attributes and hypotheses. The d-complete tree G has 2^d complete paths $\xi_1, \ldots, \xi_{2^d}$. For $i = 1, \ldots, 2^d$, we denote by a_i a solution of the equation system $\mathscr{S}(\xi_i)$. Denote $B = \{a_1, \ldots, a_{2^d}\}$. We now show that the decision tree Γ contains a complete path, which length is at least $\frac{d}{\log_2(2d)}$. We describe the process of this path construction beginning with the root of Γ.

Let the root of Γ be labeled with an attribute f_{i_0}. For $\delta \in \{0, 1\}$, we denote by B^δ the set of solutions on B of the equation system $\{f_{i_0}(x) = \delta\}$ and choose $\sigma \in \{0, 1\}$ for which $|B^\sigma| = \max\{|B^0|, |B^1|\}$. It is clear that $|B^\sigma| \geq \frac{|B|}{2} \geq \frac{|B|}{2d}$. In the considered case, the beginning of the constructed path in Γ is the root of Γ, the edge that leaves the root and is labeled with the equation system $\{f_{i_0}(x) = \sigma\}$, and the node to which this edge enters.

Let as assume now that the root of Γ is labeled with a hypothesis $H = \{f_1(x) = \delta_1, \ldots, f_n(x) = \delta_n\}$. We denote by ξ_H the complete path in G for which the system of equations $\mathcal{S}(\xi_H)$ is a subsystem of H. Let the nonterminal nodes of the complete path ξ_H be labeled with the attributes $f_{i_1}, \ldots, f_{i_d}$. For $j = 1, \ldots, d$, we denote by B_j the set of solutions on B of the equation system $\{f_{i_j}(x) = \neg\delta_{i_j}\}$. It is clear that $|B_1| + \cdots + |B_d| \geq |B| - 1$. Therefore there exists $l \in \{1, \ldots, d\}$ such that $|B_l| \geq \frac{|B|-1}{d} \geq \frac{|B|}{2d}$. In the considered case, the beginning of the constructed path in Γ is the root of Γ, the edge that leaves the root and is labeled with the equation system $\{f_{i_l}(x) = \neg\delta_{i_l}\}$, and the node to which this edge enters.

We continue the construction of the complete path in Γ in the same way such that after the tth query we will have at least $\frac{|B|}{(2d)^t}$ elements from B. The process of path construction will continue at least until $\frac{|B|}{(2d)^t} \leq 1$, i.e., at least until $\log_2 |B| \leq t \log_2(2d)$. Since $|B| = 2^d$, we have $h(\Gamma) \geq t \geq \frac{d}{\log_2(2d)}$ and $h_U^{(3)}(z) \geq \frac{d}{\log_2(2d)}$. □

Lemma 7.2 *Let $U = (A, F)$ be a binary information system, $k \in \mathbb{N} \cup \{0\}$, and U be not m-information system for $m = 0, \ldots, k$. Then there exists a $(k + 1)$-complete tree over U.*

Proof We prove the considered statement by induction on k. Let $k = 0$. In this case, U is not 0-information system. Then there exists an attribute $f \in F$, which is not constant on A. Using this attribute, it is easy to construct 1-complete tree over U.

Let the considered statement hold for some k, $k \geq 0$. We now show that it also holds for $k + 1$. Let $U = (A, F)$ be a binary information system, which is not m-information system for $m = 1, \ldots, k + 1$. Then there exists an attribute $f \in F$ such that, for any $\delta \in \{0, 1\}$, the information system $U_\delta = (A(f, \delta), F)$ is not m-information system for $m = 1, \ldots, k$. Using the inductive hypothesis, we conclude that, for any $\delta \in \{0, 1\}$, there exists a $(k + 1)$-complete tree G_δ over U_δ. Denote by G a directed tree with root in which the root is labeled with the attribute f and, for any $\delta \in \{0, 1\}$, there is an edge that leaves the root, is labeled with the equation system $\{f(x) = \delta\}$, and enters the root of the tree G_δ. One can show that the tree G is a $(k + 2)$-complete tree over U. □

Proof (of Theorem 7.2) It is clear, that $h_U^{(3)}(z) \leq h_U^{(2)}(z)$ for any problem z over U. Therefore $h_U^{(3)}(n) \leq h_U^{(2)}(n)$ for any $n \in \mathbb{N}$.

(a) Let $k \in \mathbb{N} \cup \{0\}$. We now show by induction on k that, for each binary k-information system U (not necessary infinite) for each problem z over U, the inequality $h_U^{(2)}(z) \leq k$ holds. Let $U = (A, F)$ be a binary 0-information system and z be a problem over U. Since all attributes from $F(z)$ are constant on A, the set $\Delta_U(z)$ contains only one tuple. Therefore the decision tree containing only one node labeled with this tuple solves the problem z relative to U, and $h_U^{(2)}(z) = 0$.

Let $k \in \mathbb{N} \cup \{0\}$ and, for each m, $0 \leq m \leq k$, the considered statement hold. Let us show that it holds for $k + 1$. Let $U = (A, F)$ be a binary $(k + 1)$-information system and $z = (f_1, \ldots, f_n)$ be a problem over U. For $i = 1, \ldots, n$, choose a number $\delta_i \in \{0, 1\}$ such that the information system $(A(f_i, \neg\delta_i), F)$ is m_i-information system where $1 \leq m_i \leq k$. Using the inductive hypothesis, we conclude that, for $i = 1, \ldots, n$, there is a decision tree Γ_i over z, which uses only hypotheses, solves the problem z over $(A(f_i, \neg\delta_i), F)$, and has depth at most m_i. We denote by Γ a decision tree in which the root is labeled with the hypothesis $H = \{f_1(x) = \delta_1, \ldots, f_n(x) = \delta_n\}$, the edge leaving the root and labeled with H enters the terminal node labeled with the tuple $(\delta_1, \ldots, \delta_n)$, and for $i = 1, \ldots, n$, the edge leaving the root and labeled with $\{f_i(x) = \neg\delta_i\}$ enters the root of the tree Γ_i. One can show that Γ solves the problem z relative to U and $h(\Gamma) \leq k + 1$. Therefore, $h_U^{(2)}(z) \leq k + 1$ for any problem z over U.

Let $U \in \mathscr{C}$. Then U is k-information system for some natural k and, for each problem z over U, we have $h_U^{(3)}(z) \leq h_U^{(2)}(z) \leq k$. Therefore $h_U^{(2)}(n) = O(1)$ and $h_U^{(3)}(n) = O(1)$.

(b) Let $U = (A, F) \in \mathscr{D} \setminus \mathscr{C}$. First, we show that $h_U^{(2)}(n) = O(\log n)$. Let $z = (f_1, \ldots, f_n)$ be an arbitrary problem over U. From Lemma 5.1 [9] it follows that $|\Delta_U(z)| \leq (4n)^{I(U)}$. The proof of this lemma is based on the results similar to ones obtained by Sauer [13] and Shelah [14]. We consider a decision tree Γ over z, which solves z relative to U and uses only hypotheses. This tree is constructed by the halving algorithm [1, 3]. We describe the work of this tree for an arbitrary element a from A. Set $\Delta = \Delta_U(z)$. If $|\Delta| = 1$, then the only n-tuple from Δ is the solution $z(a)$ of the problem z for the element a. Let $|\Delta| \geq 2$. For $i = 1, \ldots, m$, we denote by δ_i a number from $\{0, 1\}$ such that $|\Delta(f_i, \delta_i)| \geq |\Delta(f_i, \neg\delta_i)|$. The root of Γ is labeled with the hypothesis $H = \{f_1(x) = \delta_1, \ldots, f_n(x) = \delta_n\}$. After this query either the problem z is solved (if the answer is H) or we halve the number of objects in the set Δ (if the answer is a counterexample $\{f_i(x) = \neg\delta_i\}$). In the latter case, set $\Delta = \Delta_U(z)(f_i, \neg\delta_i)$. The decision tree Γ continues to work with the element a and the set of n-tuples Δ in the same way. Let during the work with the element a, the considered decision tree make q queries. After the $(q - 1)$th query, the number of remaining n-tuples in the set Δ is at least two and at most $(4n)^{I(U)}/2^{q-1}$. Therefore $2^q \leq (4n)^{I(U)}$ and $q \leq I(U)\log_2(4n)$. So during the processing of the element a, the decision tree Γ makes at most $I(U)\log_2(4n)$ queries. Since a is an arbitrary element from A, the depth of Γ is at most $I(U)\log_2(4n)$. Since z is an arbitrary problem over U, we obtain $h_U^{(2)}(n) = O(\log n)$. Therefore $h_U^{(3)}(n) = O(\log n)$.

Using Lemma 7.2 and the relation $U \notin \mathscr{C}$, we obtain that, for any $d \in \mathbb{N}$, there exists d-complete tree G_d over U. Let $F(G_d) = \{f_1, \ldots, f_{n_d}\}$. We know that $n_d \leq 2^d$. Denote $z_d = (f_1, \ldots, f_{n_d})$. From Lemma 7.1 it follows that $h_U^{(2)}(z_d) \geq d$ and $h_U^{(3)}(z_d) \geq \frac{d}{\log_2(2d)}$.

As a result, we have $h_U^{(2)}(2^d) \geq d$ and $h_U^{(3)}(2^d) \geq \frac{d}{\log_2(2d)}$. Let $n \in \mathbb{N}$ and $n \geq 8$. Then there exists $d \in \mathbb{N}$ such that $2^d \leq n < 2^{d+1}$. We have $d > \log_2 n - 1$, $h_U^{(2)}(n) \geq \log_2 n - 1$, $h_U^{(2)}(n) = \Omega(\log n)$, and $h_U^{(2)}(n) = \Theta(\log n)$. It is easy to show that the function $\frac{x}{\log_2(2x)}$ is nondecreasing for $x \geq 2$. Therefore $h_U^{(3)}(n) \geq \frac{\log_2 n - 1}{\log_2(2(\log_2 n - 1))}$ and $h_U^{(3)}(n) = \Omega(\frac{\log n}{\log \log n})$.

(c) Let $U = (A, F) \notin \mathscr{D}$. We now consider an arbitrary problem $z = (f_1, \ldots, f_n)$ over U and a decision tree over z which uses only hypotheses and solves the problem z over U in the following way. For a given element $a \in A$, the first query is about the hypothesis $H_1 = \{f_1(x) = 1, \ldots, f_n(x) = 1\}$. If the answer is H_1, then the problem z is solved for the element a. If, for some $i \in \{1, \ldots, n\}$, the answer is $\{f_i(x) = 0\}$, then the second query is about the hypothesis H_2 obtained from H_1 by replacing the equality $f_i(x) = 1$ with the equality $f_i(x) = 0$, etc. It is clear that after at most n queries the problem z for the element a will be solved. Thus, $h_U^{(2)}(z) \leq n$ and $h_U^{(3)}(z) \leq n$. Since z is an arbitrary problem over U, we have $h_U^{(2)}(n) \leq n$ and $h_U^{(3)}(n) \leq n$ for any $n \in \mathbb{N}$.

Let $n \in \mathbb{N}$. Since $U \notin \mathscr{D}$, there exist attributes $f_1, \ldots, f_n \in F$ such that, for any $(\delta_1, \ldots, \delta_n) \in \{0, 1\}^n$, the equation system $\{f_1(x) = \delta_1, \ldots, f_n(x) = \delta_n\}$ is consistent on A. We now consider the problem $z = (f_1, \ldots, f_n)$ and an arbitrary decision tree Γ over z, which solves the problem z over U and uses both attributes and hypotheses. Let us show that $h(\Gamma) \geq n$. If $n = 1$, then the considered inequality holds since $|\Delta_U(z)| \geq 2$. Let $n \geq 2$. It is easy to show that an equation system over z is inconsistent if and only if it contains equations $f_i(x) = 0$ and $f_i(x) = 1$ for some $i \in \{1, \ldots, n\}$. For each node v of the decision tree Γ, we denote by S_v the union of systems of equations attached to edges in the path from the root of Γ to v. A node v of Γ will be called consistent if the equation system S_v is consistent.

We now construct a complete path ξ in the decision tree Γ, which nodes are consistent. We will start constructing the path from the root that is a consistent node. Let the path reach a consistent node v of Γ. If v is a terminal node, then the path ξ is constructed. Let v be a working node labeled with an attribute $f_i \in F(z)$. Then there exists $\delta \in \{0, 1\}$ for which the system of equations $S_v \cup \{f_i(x) = \delta\}$ is consistent. Then the path ξ will pass through the edge leaving v and labeled with the system of equations $\{f_i(x) = \delta\}$. Let v be labeled with a hypothesis $H = \{f_1(x) = \delta_1, \ldots, f_n(x) = \delta_n\}$. If there exists $i \in \{1, \ldots, n\}$ such that the system of equations $S_v \cup \{f_i(x) = \neg\delta\}$ is consistent, then the path ξ will pass through the edge leaving v and labeled with the system of equations $\{f_i(x) = \neg\delta\}$. Otherwise, $S_v = H$ and the path ξ will pass through the edge leaving v and labeled with the system of equations H.

Let all edges in the path ξ be labeled with systems of equations containing one equation each. Since all nodes of ξ are consistent, the equation system $\mathscr{S}(\xi)$ is consistent. We now show that $\mathscr{S}(\xi)$ contains at least n equations. Let us assume that this system contains less than n equations. Then the set $\Delta_U(z)\pi(\xi)$ contains more than one n-tuple, which is impossible. Therefore the length of the path ξ is at least n. Let there be edges in ξ, which are labeled with hypotheses, and the first edge in ξ labeled with a hypothesis H leaves the node v. Then $S_v = H$ and the length of ξ is at least n. Therefore $h(\Gamma) \geq n$, $h_U^{(3)}(z) \geq n$, and $h_U^{(2)}(z) \geq n$. As a result, we obtain $h_U^{(3)}(n) \geq n$ and $h_U^{(2)}(n) \geq n$. Thus, $h_U^{(2)}(n) = n$ and $h_U^{(3)}(n) = n$ for any $n \in \mathbb{N}$. □

7.4 Proof of Theorem 7.3

First, we prove several auxiliary statements.

Proposition 7.1 $\mathscr{R} \subseteq \mathscr{D}$.

Proof Let $U \in \mathscr{R}$. By Theorem 7.1, $h_U^{(1)}(n) = \Theta(\log n)$. Let us assume that $U \notin \mathscr{D}$. Then, for any $n \in \mathbb{N}$, there exists a problem $z = (f_1, \ldots, f_n)$ over U such that $|\Delta_U(z)| = 2^n$. Let Γ be a decision tree over z, which solves the problem z relative to U and uses only attributes. Then Γ should have at least 2^n terminal nodes. One can show that the number of terminal nodes in the tree Γ is at most $2^{h(\Gamma)}$. Then $2^n \leq 2^{h(\Gamma)}$, $h(\Gamma) \geq n$, and $h_U(z) \geq n$. Therefore $h_U^{(1)}(n) \geq n$ for any $n \in \mathbb{N}$, which is impossible. Thus, $\mathscr{R} \subseteq \mathscr{D}$. □

Proposition 7.2 $\mathscr{C} \subseteq \mathscr{D}$.

Proof Let $U \in \mathscr{C}$. By Theorem 7.2, $h_U^{(2)}(n) = O(1)$. Let us assume that $U \notin \mathscr{D}$. Then, by Theorem 7.2, $h_U^{(2)}(n) = n$ for any $n \in \mathbb{N}$, which is impossible. Therefore $\mathscr{C} \subseteq \mathscr{D}$. □

Proposition 7.3 $\mathscr{R} \cap \mathscr{C} = \emptyset$.

Proof Assume the contrary: $\mathscr{R} \cap \mathscr{C} \neq \emptyset$ and $U = (A, F) \in \mathscr{R} \cap \mathscr{C}$. Let $r, k \in \mathbb{N}$, U be r-reduced information system, and U be k-information system. We now consider an arbitrary problem $z = (f_1, \ldots, f_n)$ over U and describe a decision tree Γ over z, which uses only attributes, solves the problem z over U, and has depth at most kr.

For $i = 1, \ldots, n$, let δ_i be a number from $\{0, 1\}$ such that $(A(f_i, \neg\delta_i), F)$ is m_i-information system with $0 \leq m_i < k$. Let t be the maximum number from the set $\{1, \ldots, n\}$ such that the system of equations $S = \{f_1(x) = \delta_1, \ldots, f_t(x) = \delta_t\}$ is consistent. Then there exists a subsystem $\{f_{i_1}(x) = \delta_{i_1}, \ldots, f_{i_p}(x) = \delta_{i_p}\}$ of the system S, which has the same set of solutions as S and for which $p \leq r$. For a given $a \in A$, the decision tree Γ computes sequentially values $f_{i_1}(a), \ldots, f_{i_p}(a)$.

If, for some $q \in \{1, \ldots, p\}$, $f_{i_1}(a) = \delta_{i_1}, \ldots, f_{i_{q-1}}(a) = \delta_{i_{q-1}}$, and $f_{i_q}(a) = \neg\delta_{i_q}$, then the decision tree Γ continues to work with the problem z and the information system $U' = (A', F)$ where A' is the set of solutions on A of the equation system $\{f_{i_1}(x) =$

Table 7.3 All 3-tuples from the set $\{0, 1\}^3$

	$\mathscr{R}$	$\mathscr{D}$	$\mathscr{C}$
1	0	0	0
2	0	1	0
3	0	1	1
4	1	1	0
5	1	0	0
6	0	0	1
7	1	0	1
8	1	1	1

$\delta_{i_1}, \ldots, f_{i_{q-1}}(x) = \delta_{i_{q-1}}, f_{i_q}(x) = \neg\delta_{i_q}\}$. We have U' is l'-information system for some $l' \leq m_{i_q} < k$.

Let $f_{i_1}(a) = \delta_{i_1}, \ldots, f_{i_p}(a) = \delta_{i_p}$. If $t = n$, then $(\delta_1, \ldots, \delta_n)$ is the solution of the problem z for the considered element a. Let $t < n$. Then the decision tree Γ continues to work with the problem z and the information system $U'' = (A'', F)$ where A'' is the set of solutions on A of the equation system $\{f_{i_1}(x) = \delta_{i_1}, \ldots, f_{i_p}(x) = \delta_{i_p}\}$. We know that the equation system $\{f_1(x) = \delta_1, \ldots, f_t(x) = \delta_t, f_{t+1}(x) = \delta_{t+1}\}$ is inconsistent. Therefore the system $\{f_{i_1}(x) = \delta_{i_1}, \ldots, f_{i_p}(x) = \delta_{i_p}, f_{t+1}(x) = \delta_{t+1}\}$ is inconsistent. Hence $A'' \subseteq A(f_{t+1}, \neg\delta_{t+1})$ and U'' is l''-information system for some $l'' \leq m_{t+1} < k$.

As a result, after computation of values of at most r attributes, we either solve the problem z or reduce the consideration of the problem z over k-information system U to the consideration of the problem z over some l-information system where $l < k$. After computation of values of at most rk attributes we solve the problem z since each problem over 0-information system has exactly one possible solution. Therefore $h_U^{(1)}(z) \leq rk$ and $h_U^{(1)}(n) = O(1)$. By Theorem 7.1, $h_U^{(1)}(n) = \Theta(\log n)$. The obtained contradiction shows that $\mathscr{R} \cap \mathscr{C} = \emptyset$. □

Proposition 7.4 *For any infinite binary information system, its indicator vector coincides with one of the rows of Table 7.1.*

Proof Table 7.3 contains as rows all 3-tuples from the set $\{0, 1\}^3$. We now show that rows with numbers 5–8 cannot be indicator vectors of infinite binary information systems. Assume the contrary: there is $i \in \{5, 6, 7, 8\}$ such that the row with the number i is the indicator vector of an infinite binary information system U. If $i = 5$, then $U \in \mathscr{R}$ and $U \notin \mathscr{D}$, but this is impossible since, by Proposition 7.1, $\mathscr{R} \subseteq \mathscr{D}$. If $i = 6$, then $U \in \mathscr{C}$ and $U \notin \mathscr{D}$, but this is impossible since, by Proposition 7.2, $\mathscr{C} \subseteq \mathscr{D}$. If $i = 7$, then $U \in \mathscr{R}$ and $U \notin \mathscr{D}$, but this is impossible since, by Proposition 7.1, $\mathscr{R} \subseteq \mathscr{D}$. If $i = 8$, then $U \in \mathscr{R}$ and $U \in \mathscr{C}$, but this is impossible since, by Proposition 7.3, $\mathscr{R} \cap \mathscr{C} = \emptyset$. Therefore, for any infinite binary information system, its indicator vector coincides with one of the rows of Table 7.3 with numbers 1–4. Thus, it coincides with one of the rows of Table 7.1. □

Define an infinite binary information system $U_1 = (A_1, F_1)$ as follows: $A_1 = \mathbb{N}$ and F_1 is the set of all functions from $\mathbb{N}$ to $\{0, 1\}$.

Lemma 7.3 *The information system U_1 belongs to the class V_1.*

Proof It is easy to show that the information system U_1 has infinite I-dimension. Therefore $U_1 \notin \mathscr{D}$. Using Proposition 7.4, we obtain $ind(U) = (0, 0, 0)$, i.e., $U_1 \in V_1$. □

For any $i \in \mathbb{N}$, we define two functions $p_i : \mathbb{N} \to \{0, 1\}$ and $l_i : \mathbb{N} \to \{0, 1\}$. Let $j \in \mathbb{N}$. Then $p_i(j) = 1$ if and only if $j = i$, and $l_i(j) = 1$ if and only if $j > i$.

Define an infinite binary information system $U_2 = (A_2, F_2)$ as follows: $A_2 = \mathbb{N}$ and $F_2 = \{p_i : i \in \mathbb{N}\} \cup \{l_i : i \in \mathbb{N}\}$.

Lemma 7.4 *The information system U_2 belongs to the class V_2.*

Proof For $n \in \mathbb{N}$, denote $S_n = \{p_1(x) = 0, \ldots, p_n(x) = 0\}$. One can show that the equation system S_n is consistent and each proper subsystem of S_n has the set of solutions different from the set of solutions of S_n. Therefore $U_2 \notin \mathscr{R}$. Using attributes from the set $\{l_i : i \in \mathbb{N}\}$, we can construct d-complete tree over U_2 for each $d \in \mathbb{N}$. By Lemma 7.1 and Theorem 7.2, $U_2 \notin \mathscr{C}$. One can show that $I(U_2) = 1$. Therefore $U_2 \in \mathscr{D}$. Thus, $ind(U_2) = (0, 1, 0)$, i.e., $U_2 \in V_2$. □

Define an infinite binary information system $U_3 = (A_3, F_3)$ as follows: $A_3 = \mathbb{N}$ and $F_3 = \{p_i : i \in \mathbb{N}\}$.

Lemma 7.5 *The information system U_3 belongs to the class V_3.*

Proof It is easy to show that U_3 is 1-information system. Therefore $U_3 \in \mathscr{C}$. Using Proposition 7.4, we obtain $ind(U_3) = (0, 1, 1)$, i.e., $U_3 \in V_3$. □

Define an infinite binary information system $U_4 = (A_4, F_4)$ as follows: $A_4 = \mathbb{N}$ and $F_4 = \{l_i : i \in \mathbb{N}\}$.

Lemma 7.6 *The information system U_4 belongs to the class V_4.*

Proof Let us consider an arbitrary consistent system of equations S over U_4. We now show that there is a subsystem of S, which has at most two equations and the same set of solutions as S. Let S contain both equations of the kind $l_i(x) = 1$ and $l_j(x) = 0$. Denote $i_0 = \max\{i : l_i(x) = 1 \in S\}$ and $j_0 = \min\{j : l_j(x) = 0 \in S\}$. One can show that the system of equations $S' = \{l_{i_0}(x) = 1, l_{j_0}(x) = 0\}$ has the same set of solutions as S. The case when S contains for some $\delta \in \{0, 1\}$ only equations of the kind $l_p(x) = \delta$ can be considered in a similar way. In this case, the equation system S' contains only one equation. Therefore the information system U_4 is 2-reduced and $U_4 \in \mathscr{R}$. Using Proposition 7.4, we obtain $ind(U_4) = (1, 1, 0)$, i.e., $U_4 \in V_4$. □

Proof (of Theorem 7.3) From Proposition 7.4 it follows that, for any infinite binary information system, its indicator vector coincides with one of the rows of Table 7.1. Using Lemmas 7.3-7.6, we conclude that each row of Table 7.1 is the indicator vector of some infinite binary information system. □

7.5 Conclusions

Based on the results of exact learning, test theory, and rough set theory, for an arbitrary infinite binary information system, we studied three functions, which characterize the dependence in the worst case of the minimum depth of a decision tree solving a problem on the number of attributes in the problem description. These three functions correspond to (1) decision trees using attributes, (2) decision trees using hypotheses, and (3) decision trees using both attributes and hypotheses. We described possible types of behavior for each of these three functions. We also studied join behavior of these functions and distinguished four complexity classes of infinite binary information systems.

References

1. Angluin, D.: Queries and concept learning. Mach. Learn. **2**(4), 319–342 (1988)
2. Angluin, D.: Queries revisited. Theor. Comput. Sci. **313**(2), 175–194 (2004)
3. Littlestone, N.: Learning quickly when irrelevant attributes abound: a new linear-threshold algorithm. Mach. Learn. **2**(4), 285–318 (1988)
4. Maass, W., Turán, G.: Lower bound methods and separation results for on-line learning models. Mach. Learn. **9**, 107–145 (1992)
5. Moshkov, M.: On conditional tests. Sov. Phys. Dokl. **27**, 528–530 (1982)
6. Moshkov, M.: Conditional tests. In: Yablonskii, S.V. (ed.) Problemy Kibernetiki, vol. 40, pp. 131–170. Nauka Publishers, Moscow (1983). (in Russian)
7. Moshkov, M.: On depth of conditional tests for tables from closed classes. In: Markov, A.A. (ed.) Combinatorial-Algebraic and Probabilistic Methods of Discrete Analysis, pp. 78–86. Gorky University Press, Gorky (1989). (in Russian)
8. Moshkov, M.: Test theory and problems of machine learning. In: International School-Seminar on Discrete Mathematics and Mathematical Cybernetics, Ratmino, Russia, 31 May–3 June 2001, pp. 6–10. MAX Press, Moscow (2001)

9. Moshkov, M.: Time complexity of decision trees. In: Peters, J.F., Skowron, A. (eds.) Trans. Rough Sets III, Lecture Notes in Computer Science, vol. 3400, pp. 244–459. Springer (2005)
10. Moshkov, M.: On the depth of decision trees with hypotheses. Entropy **24**(1), 116 (2022). https://doi.org/10.3390/e24010116
11. Naiman, D.Q., Wynn, H.P.: Independence number and the complexity of families of sets. Discrete Math. **154**, 203–216 (1996)
12. Pawlak, Z.: Information systems theoretical foundations. Inf. Syst. **6**(3), 205–218 (1981)
13. Sauer, N.: On the density of families of sets. J. Comb. Theory (A) **13**, 145–147 (1972)
14. Shelah, S.: A combinatorial problem; stability and order for models and theories in infinitary languages. Pac. J. Math. **41**, 241–261 (1972)

8 Infinite Binary Information Systems. Decision Trees of Types 4 and 5

In this chapter, which is based on [4, 6], we continue to study decision trees with hypotheses over infinite binary information systems. In Chap. 7, we investigated arbitrary hypotheses (an analog of equivalence queries from exact learning). The present chapter is devoted to the study of proper hypotheses (an analog of proper equivalence queries).

We consider infinite binary information systems each of which consists of an infinite set of elements A and an infinite set F of functions (attributes) from A to $\{0, 1\}$. We define the notion of a problem described by a finite number of attributes $f_1, \ldots, f_n$ from F: for a given element $a \in A$, we should recognize the tuple $(f_1(a), \ldots, f_n(a))$. To this end, we can use decision trees based on two types of queries. We can ask about the value of an attribute $f_i \in \{f_1, \ldots, f_n\}$. We will obtain an answer of the kind $f_i(x) = \delta$, where $\delta \in \{0, 1\}$. We can also ask if a hypothesis $f_1(x) = \delta_1, \ldots, f_n(x) = \delta_n$ is true, where $\delta_1, \ldots, \delta_n \in \{0, 1\}$. Either this hypothesis will be confirmed or we will obtain a counterexample in the form $f_i(x) = \neg\delta_i$. The considered hypothesis is called proper if there exists an element $a \in A$ such that $f_1(a) = \delta_1, \ldots, f_n(a) = \delta_n$.

Based on the results of exact learning, test theory, and rough set theory, for an arbitrary infinite binary information system, we studied in Chap. 7 three functions of Shannon type, which characterize the dependence in the worst case of the minimum depth of a decision tree solving a problem on the number of attributes in the problem description. The considered three functions correspond to the following three types of decision trees:

(1) Only attributes are used in decision trees.
(2) Only hypotheses are used in decision trees.
(3) Both attributes and hypotheses are used in decision trees.

We showed that the first function has two possible types of behavior: logarithmic and linear. The second and the third functions have three possible types of behavior: constant,

M. Azad et al., *Decision Trees with Hypotheses*, Synthesis Lectures on Intelligent Technologies, https://doi.org/10.1007/978-3-031-08585-7_8

logarithmic, and linear. We also studied joint behavior of these three functions and described four complexity classes of infinite binary information systems.

In the present chapter, we study two functions of Shannon type, which characterize the dependence in the worst case of the minimum depth of a decision tree solving a problem on the number of attributes in the problem description. These functions correspond to the following two types of decision trees:

(4) Only proper hypotheses are used in decision trees.
(5) Both attributes and proper hypotheses are used in decision trees.

We prove that both functions have three possible types of behavior: constant, logarithmic, and linear. Results related to these functions were presented in [4] without proofs. We also study joint behavior of all five functions of Shannon type and describe seven complexity classes of infinite binary information systems.

The rest of the chapter is organized as follows. In Sect. 8.1, we define five functions of Shannon type and, in Sect. 8.2, we present the main results of this chapter. Sections 8.3 and 8.4 contain proofs of the main results, and Sect. 8.5—short conclusions.

8.1 Five Functions of Shannon Type

Let $U = (A, F)$ be an infinite binary information system and $z = (f_1, \ldots, f_n)$ be a problem over U. A hypothesis $H = \{f_1(x) = \delta_1, \ldots, f_n(x) = \delta_n\}$ over z is called *proper* (for U) if the system of equations H is consistent on A.

We denote by $h_U^{(1)}(z)$ the minimum depth of a decision tree over z, which solves z relative to U and uses only attributes from $F(z)$. We denote by $h_U^{(2)}(z)$ the minimum depth of a decision tree over z, which solves z relative to U and uses only hypotheses over z. We denote by $h_U^{(3)}(z)$ the minimum depth of a decision tree over z, which solves z relative to U and uses both attributes from $F(z)$ and hypotheses over z. We denote by $h_U^{(4)}(z)$ the minimum depth of a decision tree over z, which solves z relative to U and uses only proper hypotheses over z. We denote by $h_U^{(5)}(z)$ the minimum depth of a decision tree over z, which solves z relative to U and uses both attributes from $F(z)$ and proper hypotheses over z.

For $i = 1, 2, 3, 4, 5$, we define a function of Shannon type $h_U^{(i)}(n)$ that characterizes dependence of $h_U^{(i)}(z)$ on $\dim z$ in the worst case. Let $i \in \{1, 2, 3, 4, 5\}$ and $n \in \mathbb{N}$. Then

$$h_U^{(i)}(n) = \max\{h_U^{(i)}(z) : z \in P(U), \dim z \leq n\} .$$

8.2 Main Results

In this section, we consider the main results of this chapter.

Let $U = (A, F)$ be a binary information system and $r \in \mathbb{N}$. We will say that the information system U is *r-i-reduced* if, for each inconsistent on A system of equations over U, there exists a subsystem of this system that is inconsistent and contains at most r equations. We denote by $\mathscr{I}$ the class of infinite binary information systems each of which is r-i-reduced for some $r \in \mathbb{N}$.

We proved in Chap. 7 that $\mathscr{C} \subseteq \mathscr{D}$ – see Proposition 7.2. Therefore, for any infinite binary information system U, either $U \in \mathscr{C} \cap \mathscr{I}$, or $U \in (\mathscr{D} \setminus \mathscr{C}) \cap \mathscr{I}$, or $U \in \mathscr{D} \setminus \mathscr{I}$, or $U \notin \mathscr{D}$. The following theorem was presented in [4] without proof.

Theorem 8.1 *Let U be an infinite binary information system. Then the following statements hold:*

(a) If $U \in \mathscr{C} \cap \mathscr{I}$, *then* $h_U^{(4)}(n) = O(1)$ *and* $h_U^{(5)}(n) = O(1)$.

(b) If $U \in (\mathscr{D} \setminus \mathscr{C}) \cap \mathscr{I}$, *then* $h_U^{(4)}(n) = \Theta(\log n)$, $h_U^{(5)}(n) = \Omega(\frac{\log n}{\log \log n})$, *and* $h_U^{(5)}(n) = O(\log n)$.

(c) If $U \in \mathscr{D} \setminus \mathscr{I}$ *and* $i \in \{4, 5\}$, *then* $h_U^{(i)}(n) \geq n - 1$ *for infinitely many* $n \in \mathbb{N}$ *and* $h_U^{(i)}(n) \leq n$ *for any* $n \in \mathbb{N}$.

(d) If $U \notin \mathscr{D}$, *then* $h_U^{(4)}(n) = n$ *and* $h_U^{(5)}(n) = n$ *for any* $n \in \mathbb{N}$.

Let U be an infinite binary information system. We now consider the joint behavior of the functions $h_U^{(1)}(n)$, $h_U^{(2)}(n)$, $h_U^{(3)}(n)$, $h_U^{(4)}(n)$, and $h_U^{(5)}(n)$. It depends on the belonging of the information system U to the classes $\mathscr{R}, \mathscr{D}, \mathscr{C}$, and $\mathscr{I}$. We correspond to the information system U its *extended indicator vector* $eind(U) = (c_1, c_2, c_3, c_4) \in \{0, 1\}^4$ in which $c_1 = 1$ if and only if $U \in \mathscr{R}$, $c_2 = 1$ if and only if $U \in \mathscr{D}$, $c_3 = 1$ if and only if $U \in \mathscr{C}$, and $c_4 = 1$ if and only if $U \in \mathscr{I}$.

Theorem 8.2 *For any infinite binary information system, its extended indicator vector coincides with one of the rows of Table 8.1. Each row of Table 8.1 is the extended indicator vector of some infinite binary information system.*

For $i = 1, \ldots, 7$, we denote by $\mathscr{V}_i$ the class of all infinite binary information systems, which extended indicator vector coincides with the ith row of Table 8.1. Table 8.2 summarizes Theorems 7.1, 7.2, 8.1, and 8.2. The first column contains the name of *complexity class* $\mathscr{V}_i$. The next four columns describe the extended indicator vector of information systems from this class. The last five columns $h_U^{(1)}(n)$, ..., $h_U^{(5)}(n)$ contain information about behavior of the functions $h_U^{(1)}(n)$, ..., $h_U^{(5)}(n)$ for information systems from the class $\mathscr{V}_i$. The notation $\approx \log n$ in a column $h_U^{(i)}(n)$ means that $h_U^{(i)}(n) = \Omega(\frac{\log n}{\log \log n})$ and $h_U^{(i)}(n) = O(\log n)$. The

Table 8.1 Possible extended indicator vectors of infinite binary information systems

	$\mathscr{R}$	$\mathscr{D}$	$\mathscr{C}$	$\mathscr{I}$
1	0	0	0	0
2	0	0	0	1
3	0	1	0	0
4	0	1	0	1
5	0	1	1	0
6	0	1	1	1
7	1	1	0	1

Table 8.2 Summary of Theorems 7.1, 7.2, 8.1, and 8.2

	$\mathscr{R}$	$\mathscr{D}$	$\mathscr{C}$	$\mathscr{I}$	$h_U^{(1)}(n)$	$h_U^{(2)}(n)$	$h_U^{(3)}(n)$	$h_U^{(4)}(n)$	$h_U^{(5)}(n)$
$\mathscr{V}_1$	0	0	0	0	n	n	n	n	n
$\mathscr{V}_2$	0	0	0	1	n	n	n	n	n
$\mathscr{V}_3$	0	1	0	0	n	$\Theta(\log n)$	$\approx \log n$	$\approx n$	$\approx n$
$\mathscr{V}_4$	0	1	0	1	n	$\Theta(\log n)$	$\approx \log n$	$\Theta(\log n)$	$\approx \log n$
$\mathscr{V}_5$	0	1	1	0	n	$O(1)$	$O(1)$	$\approx n$	$\approx n$
$\mathscr{V}_6$	0	1	1	1	n	$O(1)$	$O(1)$	$O(1)$	$O(1)$
$\mathscr{V}_7$	1	1	0	1	$\Theta(\log n)$	$\Theta(\log n)$	$\approx \log n$	$\Theta(\log n)$	$\approx \log n$

notation $\approx n$ in a column $h_U^{(i)}(n)$ means that $h_U^{(i)}(n) \leq n$ for any $n \in \mathbb{N}$ and $h_U^{(i)}(n) \geq n - 1$ for infinitely many $n \in \mathbb{N}$.

Note that it is possible to consider the union $\mathscr{V}_1 \cup \mathscr{V}_2$ of the complexity classes $\mathscr{V}_1$ and $\mathscr{V}_2$ as one complexity class since the functions $h_U^{(1)}(n), \ldots, h_U^{(5)}(n)$ have the same behavior for information systems from these classes. However, in this chapter, we study $\mathscr{V}_1$ and $\mathscr{V}_2$ as different complexity classes.

8.3 Proof of Theorem 8.1

In this section, we prove Theorem 8.1. First, we consider several auxiliary statements.

From Proposition 7.2 it follows that $\mathscr{C} \subseteq \mathscr{D}$. Using this fact, one can show that, for any infinite binary information system U, either $U \in \mathscr{C} \cap \mathscr{I}$, or $U \in (\mathscr{D} \setminus \mathscr{C}) \cap \mathscr{I}$, or $U \in \mathscr{D} \setminus \mathscr{I}$, or $U \notin \mathscr{D}$.

Lemma 8.1 *Let $U = (A, F)$ be a binary information system, z be a problem over U, and Γ_1 be a decision tree over z that solves the problem z relative to U and uses both attributes* *from $F(z)$ and proper hypotheses over z. Then there exists a decision tree Γ_2 over z that solves the problem z relative to U, uses only proper hypotheses over z, and satisfies the inequality $h(\Gamma_2) \leq 2^{h(\Gamma_1)} - 1$.*

Proof We prove this statement by the induction on the depth $h(\Gamma_1)$ of the decision tree Γ_1. Let $h(\Gamma_1) = 0$. Then, as the decision tree Γ_2, we can take the decision tree Γ_1. It is clear that $h(\Gamma_2) = 2^{h(\Gamma_1)} - 1$. We now assume that the considered statement is true for any binary information system, any problem over this system, and any decision tree over the considered problem that solves this problem, uses both attributes and proper hypotheses, and has depth at most k, $k \geq 0$.

Let $U = (A, F)$ be a binary information system, $z = (f_1, \ldots, f_n)$ be a problem over U, and Γ_1 be a decision tree over z that solves the problem z relative to U, uses both attributes from $F(z)$ and proper hypotheses over z, and satisfies the condition $h(\Gamma_1) = k + 1$. We now show that there exists a decision tree Γ_2 over z, which solves the problem z relative to U, uses only proper hypotheses over z, and which depth is at most $2^{k+1} - 1$.

Let the root of Γ_1 be labeled with a proper hypothesis

$$H = \{f_1(x) = \delta_1, \ldots, f_n(x) = \delta_n\} .$$

Then there are $n + 1$ edges, which leave the root, are labeled with the systems of equations $H, \{f_1(x) = \neg\delta_1\}, \ldots, \{f_n(x) = \neg\delta_n\}$, and enter the roots of subtrees $G_0, G_1, \ldots, G_n$ of the tree Γ_1, respectively. It is clear that, for $i = 1, \ldots, n$, G_i is a decision tree over z, which solves the problem z relative to the binary information system $U_i = (A(f_i, \neg\delta_i), F)$, uses only attributes and proper hypotheses for U_i, and satisfies the inequality $h(G_i) \leq k$. Using the inductive hypothesis, we obtain that, for $i = 1, \ldots, n$, there exists a decision tree G_i' over z that solves the problem z relative to U_i, uses only proper hypotheses for U_i, and satisfies the inequalities $h(G_i') \leq 2^{h(G_i)} - 1 \leq 2^k - 1$. Let G_0' be the decision tree, which contains only one node labeled with the tuple $(\delta_1, \ldots, \delta_n)$. We denote by Γ_2 the decision tree over z that is obtained from the decision tree Γ_1 by replacing the subtrees $G_0, G_1, \ldots, G_n$ with the subtrees $G_0', G_1', \ldots, G_n'$. It is easy to show that Γ_2 is a decision tree over z, which solves the problem z relative to U, uses only proper hypotheses for U, and satisfies the inequalities $h(\Gamma_2) \leq 2^k - 1 + 1 \leq 2^{h(\Gamma_1)} - 1$.

Let the root of Γ_1 be labeled with an attribute f_i. Then there are two edges, which leave the root, are labeled with the systems of equations $\{f_i(x) = 0\}$ and $\{f_i(x) = 1\}$, and enter the roots of subtrees T_0 and T_1 of the tree Γ_1, respectively. It is clear that, for $p = 0, 1$, T_p is a decision tree over z, which solves the problem z relative to the binary information system $U_p = (A(f_i, p), F)$, uses only attributes and proper hypotheses for U_p, and satisfies the inequality $h(T_p) \leq k$. Using the inductive hypothesis, we obtain that, for $p = 0, 1$, there exists a decision tree T_p' over z that solves the problem z relative to U_p, uses only proper hypotheses for U_p, and satisfies the inequalities $h(T_p') \leq 2^{h(T_p)} - 1 \leq 2^k - 1$. We denote by T the decision tree obtained from the decision tree T_0' by replacing each terminal node of T_0' with the decision tree T_1'.

Denote by Γ_2 the decision tree obtained from T by the following transformation of each complete path ξ in T. If $\mathscr{A}(\xi) = \emptyset$, then we keep the path ξ untouched. Let $\mathscr{A}(\xi) \neq \emptyset$, $\bar{\delta} = (\delta_1, \ldots, \delta_n)$ be the tuple that was attached to the terminal node of the tree T_0' through which the path ξ passes, and $\bar{\sigma} = (\sigma_1, \ldots, \sigma_n)$ be the tuple attached to the terminal node of ξ. Since $\mathscr{A}(\xi) \neq \emptyset$, at least one of the tuples $\bar{\delta}$ and $\bar{\sigma}$ belongs to the set $\Delta_U(z)$. Let, for the definiteness, $\bar{\delta} \in \Delta_U(z)$. Denote $H = \{f_1(x) = \delta_1, \ldots, f_n(x) = \delta_n\}$. We replace the terminal node of the path ξ with the working node labeled with the hypothesis H, which is proper for U. There are $n + 1$ edges that leave this node and are labeled with the systems of equations $H, \{f_1(x) = \neg\delta_1\}, \ldots, \{f_n(x) = \neg\delta_n\}$, respectively. The edge labeled with H enters to the terminal node labeled with the tuple $\bar{\delta}$. All other edges enter to terminal nodes labeled with the tuple $\bar{\sigma}$. One can show that Γ_2 is a decision tree over z that solves the problem z relative to U, uses only proper hypotheses for U, and satisfies the relations $h(\Gamma_2) \leq 2(2^k - 1) + 1 = 2^{h(\Gamma_1)} - 1$. □

Lemma 8.2 *Let $U = (A, F)$ be an infinite binary information system. Then $h_U^{(3)}(n) \leq h_U^{(5)}(n) \leq h_U^{(4)}(n) \leq n$ and $h_U^{(2)}(n) \leq h_U^{(4)}(n)$ for any $n \in \mathbb{N}$.*

Proof It is clear, that $h_U^{(3)}(z) \leq h_U^{(5)}(z) \leq h_U^{(4)}(z)$ and $h_U^{(2)}(z) \leq h_U^{(4)}(z)$ for any problem z over U. Therefore $h_U^{(3)}(n) \leq h_U^{(5)}(n) \leq h_U^{(4)}(n)$ and $h_U^{(2)}(n) \leq h_U^{(4)}(n)$ for any $n \in \mathbb{N}$.

We now consider an arbitrary problem $z = (f_1, \ldots, f_n)$ over U and a decision tree over z, which uses only proper hypotheses for U and solves the problem z relative to U in the following way. For a given element $a \in A$, the first query is about an arbitrary proper hypothesis $H_1 = \{f_1(x) = \delta_1, \ldots, f_n(x) = \delta_n\}$ for U. If the answer is H_1, then the problem z is solved for the element a. If, for some $i \in \{1, \ldots, n\}$, the answer is $\{f_i(x) = \neg\delta_i\}$, then the second query is about a proper hypothesis $H_2 = \{f_1(x) = \sigma_1, \ldots, f_n(x) = \sigma_n\}$ such that $\sigma_i = \neg\delta_i$. If the answer is H_2, then the problem z is solved for the element a. If, for some $j \in \{1, \ldots, n\}$, the answer is $\{f_j(x) = \neg\sigma_j\}$, then the third query is about a proper hypothesis $H_3 = \{f_1(x) = \gamma_1, \ldots, f_n(x) = \gamma_n\}$ such that $\gamma_i = \neg\delta_i$ and $\gamma_j = \neg\sigma_j$, etc. It is clear that after at most n queries the problem z for the element a will be solved. Thus, $h_U^{(4)}(z) \leq n$. Since z is an arbitrary problem over U, we have $h_U^{(4)}(n) \leq n$ for any $n \in \mathbb{N}$. □

Proof (of Theorem** 8.1**) (a) Let $r \in \mathbb{N}$. We now show by induction on $k \in \mathbb{N} \cup \{0\}$ that, for each binary r-i-reduced k-information system U (not necessary infinite) for each problem z over U, the inequality $h_U^{(5)}(z) \leq rk$ holds.

Let $U = (A, F)$ be a binary r-i-reduced 0-information system and z be a problem over U. Since all attributes from $F(z)$ are constant on A, the set $\Delta_U(z)$ contains only one tuple. Therefore the decision tree consisting of one node labeled with this tuple solves the problem z relative to U, and $h_U^{(5)}(z) = 0$.

Let $k \in \mathbb{N} \cup \{0\}$ and, for each m, $0 \leq m \leq k$, the considered statement hold. Let us show that it holds for $k + 1$. Let $U = (A, F)$ be a binary r-i-reduced $(k + 1)$-information system and $z = (f_1, \ldots, f_n)$ be a problem over U. For $i = 1, \ldots, n$, choose a number $\delta_i \in \{0, 1\}$ such that the information system $(A(f_i, \neg\delta_i), F)$ is m_i-information system where $1 \leq m_i \leq k$. It is easy to show that $(A(f_i, \neg\delta_i), F)$ is r-i-reduced information system. Using the inductive hypothesis, we conclude that, for $i = 1, \ldots, n$, there is a decision tree Γ_i over z, which uses both attributes from $F(z)$ and proper hypotheses for $(A(f_i, \neg\delta_i), F)$, solves the problem z relative to $(A(f_i, \neg\delta_i), F)$, and has depth at most rm_i.

Let the hypothesis $H = \{f_1(x) = \delta_1, \ldots, f_n(x) = \delta_n\}$ be proper for U. We denote by T_1 a decision tree in which the root is labeled with the hypothesis H, the edge leaving the root and labeled with H enters the terminal node labeled with the tuple $(\delta_1, \ldots, \delta_n)$, and for $i = 1, \ldots, n$, the edge leaving the root and labeled with $\{f_i(x) = \neg\delta_i\}$ enters the root of the tree Γ_i. One can show that T_1 is a decision tree over z, which uses both attributes and proper hypotheses for U, solves the problem z relative to U, and satisfies the inequalities $h(T_1) \leq rk + 1 \leq r(k+1)$.

Let the hypothesis H be not proper for U. Then the equation system $\{f_1(x) = \delta_1, \ldots, f_n(x) = \delta_n\}$ is inconsistent on A, and there exists its subsystem $\{f_{i_1}(x) = \delta_{i_1}, \ldots, f_{i_t}(x) = \delta_{i_t}\}$, which is inconsistent on A and for which $t \leq r$. We denote by G a decision tree over z with 2^t terminal nodes in which each terminal node is labeled with n-tuple $(0, \ldots, 0)$, and each complete path contains t working nodes labeled with attributes $f_{i_1}, \ldots, f_{i_t}$ starting from the root. We denote by T_2 a decision tree obtained from the decision tree G by transformation of each complete path ξ in G. Let $\{f_{i_1}(x) = \sigma_1\}, \ldots, \{f_{i_t}(x) = \sigma_t\}$ be equation systems attached to edges leaving the working nodes of ξ labeled with the attributes $f_{i_1}, \ldots, f_{i_t}$, respectively. If $(\sigma_1, \ldots, \sigma_t) = (\delta_{i_1}, \ldots, \delta_{i_t})$, then we keep the path ξ untouched. Otherwise, let j be the minimum number from the set $\{1, \ldots, t\}$ such that $\sigma_j = \neg\delta_{i_j}$. In this case, we replace the terminal node of the path ξ with the root of the decision tree Γ_{i_j}. One can show that T_2 is a decision tree over z, which uses both attributes and proper hypotheses for U, solves the problem z relative to U, and satisfies the inequalities $h(T_2) \leq rk + t \leq r(k+1)$. Therefore, $h_U^{(5)}(z) \leq r(k+1)$ for any problem z over U.

Let $U \in \mathscr{C} \cap \mathscr{I}$. Then U is r-i-reduced k-information system for some natural r and k, and $h_U^{(5)}(z) \leq rk$ for each problem z over U. From Lemma 8.1 it follows that $h_U^{(4)}(z) \leq 2^{rk} - 1$ for each problem z over U. Therefore $h_U^{(4)}(n) = O(1)$ and $h_U^{(5)}(n) = O(1)$.

(b) Let $U = (A, F) \in (\mathscr{D} \setminus \mathscr{C}) \cap \mathscr{I}$. By Lemma 8.2, $h_U^{(5)}(n) \geq h_U^{(3)}(n)$ and $h_U^{(4)}(n) \geq h_U^{(2)}(n)$ for any $n \in \mathbb{N}$. Using the fact that $U \in \mathscr{D} \setminus \mathscr{C}$ and Theorem 7.2, we obtain $h_U^{(2)}(n) = \Omega(\log n)$ and $h_U^{(3)}(n) = \Omega(\frac{\log n}{\log\log n})$. Therefore $h_U^{(4)}(n) = \Omega(\log n)$ and $h_U^{(5)}(n) = \Omega(\frac{\log n}{\log\log n})$.

Since the information systems U belongs to the set $\mathscr{D}$, it has finite I-dimension $I(U)$. Since $U \in \mathscr{I}$, the information system U is r-i-reduced for some natural r. We assume that $r \geq 2$. We can do it because each t-i-reduced information system, $t \in \mathbb{N}$, is $(t+1)$-i-reduced.

We now show that $h_U^{(4)}(n) = O(\log n)$. Let $z = (f_1, \ldots, f_n)$ be an arbitrary problem over U. From Lemma 5.1 [5] it follows that $|\Delta_U(z)| \leq (4n)^{I(U)}$. The proof of this lemma is based on the results similar to ones obtained by Sauer [7] and Shelah [8].

We consider a decision tree Γ over z, which solves the problem z relative to U and uses only proper hypotheses for U. This tree is constructed by a variant of the halving algorithm [1–3]. We describe the work of this tree for an arbitrary element a from A. Set

$\Delta = \Delta_U(z)$. If $|\Delta| = 1$, then the only n-tuple from Δ is the solution $z(a)$ to the problem z for the element a. Let $|\Delta| \geq 2$. For $i = 1, \ldots, n$, we denote by δ_i a number from $\{0, 1\}$ such that $|\Delta(f_i, \delta_i)| \geq |\Delta(f_i, \neg\delta_i)|$.

Let the system of equations $H = \{f_1(x) = \delta_1, \ldots, f_n(x) = \delta_n\}$ be consistent on A. In this case, the root of Γ is labeled with the proper hypothesis H. After this query, either the problem z will be solved (if the answer is H) or the number of remaining tuples in Δ will be at most $|\Delta|/2$ (if the answer is a counterexample $\{f_i(x) = \neg\delta_i\}$).

Let the system of equations H be inconsistent on A. For any inconsistent subsystem B of H, there exists a subsystem C of B, which is inconsistent and contains at most r equations. Then the system C contains at least one equation $f_i(x) = \delta_i$ such that $|\Delta(f_i, \neg\delta_i)| \geq |\Delta|/r$. If we assume the contrary, we obtain that the system C is consistent, which is impossible. Let $f_i \in \{f_1, \ldots, f_n\}$. The attribute f_i is called balanced if $|\Delta(f_i, \neg\delta_i)| \geq |\Delta|/r$, and unbalanced if $|\Delta(f_i, \neg\delta_i)| < |\Delta|/r$.

We denote by H_u the subsystem of H consisting of all equations $f_i(x) = \delta_i$ from H with unbalanced attributes f_i. We now show that the system H_u is consistent. Let us assume the contrary. Then it will contain at least one equation for balanced attribute, which is impossible. Let a be a solution from A to the system H_u, and $f_1(a) = \sigma_1, \ldots, f_n(a) = \sigma_n$. Then the system of equations $P = \{f_1(x) = \sigma_1, \ldots, f_n(x) = \sigma_n\}$ is consistent on A.

In the considered case, the root of Γ is labeled with the proper hypothesis P. After this query, either the problem z will be solved (if the answer is P), or the number of remaining tuples in Δ will be less than $|\Delta|/r$ (if the answer is a counterexample $\{f_i(x) = \neg\sigma_i\}$ and f_i is an unbalanced attribute), or the number of remaining tuples in Δ will be at most $|\Delta|/2$ (if the answer is a counterexample $\{f_i(x) = \neg\sigma_i\}$, $\sigma_i = \delta_i$, and f_i is a balanced attribute), or the number of remaining tuples in Δ will be at most $|\Delta|(1 - 1/r)$ (if the answer is a counterexample $\{f_i(x) = \neg\sigma_i\}$, $\sigma_i = \neg\delta_i$, and f_i is a balanced attribute).

After the first query (H or P) of the decision tree Γ, either the problem z will be solved or the number of remaining tuples in Δ will be at most $|\Delta|(1 - 1/r)$. In the latter case, when the answer is a counterexample of the kind $\{f_i(x) = \neg\gamma_i\}$ ($\gamma_i = \delta_i$ if the first query is H and $\gamma_i = \sigma_i$ if the first query is P) set $\Delta = \Delta_U(z)(f_i, \neg\gamma_i)$. It is easy to show that the information system $(A(f_i, \neg\gamma_i), F)$ is also r-i-reduced. The decision tree Γ continues to work with the element a and the set of n-tuples Δ in the same way.

Let during the work with the element a, the decision tree Γ make q queries. After the $(q - 1)$th query, the number of remaining n-tuples in the set Δ is at least two and at most $(4n)^{I(U)}(1 - 1/r)^{q-1}$. Therefore $(1 + 1/(r - 1))^q \leq (4n)^{I(U)}$ and $q \ln(1 + 1/(r - 1)) \leq I(U) \ln(4n)$. Taking into account that $\ln(1 + 1/m) > 1/(m + 1)$ for any natural m, we obtain $q < rI(U) \ln(4n)$. So during the processing of the element a, the decision tree Γ makes at most $rI(U) \ln(4n)$ queries. Since a is an arbitrary element from A, the depth of Γ is at most $rI(U) \ln(4n)$ and $h_U^{(4)}(z) \leq rI(U) \ln(4n)$. Since z is an arbitrary problem over U, we obtain $h_U^{(4)}(n) = O(\log n)$. By Lemma 8.2, $h_U^{(5)}(n) = O(\log n)$.

(c) Let $U = (A, F) \in \mathscr{D} \setminus \mathscr{I}$. From Lemma 8.2 it follows that $h_U^{(5)}(n) \leq h_U^{(4)}(n) \leq n$ for any $n \in \mathbb{N}$. We now show that, for any $m \in \mathbb{N}$, there exists a natural n such that $n \geq m$, $h_U^{(4)}(n) \geq n - 1$, and $h_U^{(5)}(n) \geq n - 1$.

Let $m \in \mathbb{N}$. Since $U \notin \mathscr{I}$, there exists a system of equations P over U with $n \geq m$ equations such that P is inconsistent but each proper subsystem of P is consistent on A. Let, for the definiteness, $P = \{f_1(x) = 0, \ldots, f_n(x) = 0\}$. Consider the problem $z = (f_1, \ldots, f_n)$ over U. Then, for $i = 1, \ldots, n$, the set $\Delta_U(z)$ contains n-tuple $\bar{\delta}_i = (0, \ldots, 0, 1, 0, \ldots, 0)$ in which all digits with the exception of the ith one are equal to 0.

Let Γ be a decision tree over z that solves the problem z relative to U and uses both attributes and proper hypotheses for U. We consider a complete path ξ in Γ for which $\pi(\xi) = (f_{i_1}, 0) \cdots (f_{i_t}, 0)$. Such complete path exists since P is not a proper hypothesis. Since Γ solves the problem z, the set $\Delta_U(z)\pi(\xi)$ contains at most one tuple. If we assume that $t < n - 1$, we obtain that $\Delta_U(z)\pi(\xi)$ contains at least two tuples. Therefore $t \geq n - 1$ and $h(\Gamma) \geq n - 1$. Thus, $h_U^{(5)}(z) \geq n - 1$, $h_U^{(5)}(n) \geq n - 1$ and, by Lemma 8.2, $h_U^{(4)}(n) \geq n - 1$.

(d) Let $U \notin \mathscr{D}$. From Lemma 8.2 it follows that $h_U^{(3)}(n) \leq h_U^{(5)}(n) \leq h_U^{(4)}(n) \leq n$ for any $n \in \mathbb{N}$. By Theorem 7.2, $h_U^{(3)}(n) = n$ for any $n \in \mathbb{N}$. Thus, $h_U^{(5)}(n) = h_U^{(4)}(n) = n$ for any $n \in \mathbb{N}$. □

8.4 Proof of Theorem 8.2

First, we consider several auxiliary statements.

Lemma 8.3 $\mathscr{R} \subseteq \mathscr{I}$.

Proof Let $U = (A, F) \in \mathscr{R}$. Then U is r-restricted for some natural r. We now show that U is $(r + 1)$-i-restricted. Let S be an arbitrary inconsistent on A equation system over U and S' be a subsystem of S with the maximum number of equations that is consistent. Since U is r-restricted, the system S' has a subsystem S'' with at most r equations and the same set of solutions on A as the system S'. It is clear that there exists an equation $f(x) = \delta$ from S such that the system of equations $S' \cup \{f(x) = \delta\}$ is inconsistent. Then the subsystem $S'' \cup \{f(x) = \delta\}$ of S with at most $r + 1$ equations is inconsistent. Therefore U is $(r + 1)$-i-restricted and $U \in \mathscr{I}$. □

Lemma 8.4 *For any infinite binary information system, its extended indicator vector coincides with one of the rows of Table 8.1.*

Proof Let U be an infinite binary information system and

$$eind(U) = (c_1, c_2, c_3, c_4) .$$

Table 8.3 All extensions of rows of Table 7.1

	$\mathscr{R}$	$\mathscr{D}$	$\mathscr{C}$	$\mathscr{I}$
1	0	0	0	0
2	0	0	0	1
3	0	1	0	0
4	0	1	0	1
5	0	1	1	0
6	0	1	1	1
7	1	1	0	1
8	1	1	0	0

Then $ind(U) = (c_1, c_2, c_3)$. By Theorem 7.3, (c_1, c_2, c_3) is a row of Table 7.1. Therefore, for each infinite binary information system, its extended indicator vector is an extension of a row of Table 7.1: it can be obtained from the row by adding the fourth digit, which is equal to 0 or 1. Table 8.3 contains all extensions of rows of Table 7.1. We now show that the row with number 8 cannot be the extended indicator vector of an infinite binary information system. Assume the contrary: there is an infinite binary information system U' such that $eind(U') = (1, 1, 0, 0)$. Then $U' \in \mathscr{R}$ and $U' \notin \mathscr{I}$, but this is impossible since, by Lemma 8.3, $\mathscr{R} \subseteq \mathscr{I}$. Therefore, for any infinite binary information system, its extended indicator vector coincides with one of the rows of Table 8.3 with numbers 1-7. Thus, it coincides with one of the rows of Table 8.1. □

We now define seven infinite binary information systems $U'_1, \ldots, U'_7$ and prove that these systems belong to the complexity classes $\mathscr{V}_1, \ldots, \mathscr{V}_7$, respectively. Four of these systems were considered in Chap. 7: $U'_1 = U_1$, $U'_3 = U_2$, $U'_6 = U_3$, and $U'_7 = U_4$. For the completeness, we repeat some reasonings from this chapter.

Define an infinite binary information system $U'_1 = (A_1, F_1)$ as follows: $A_1 = \mathbb{N}$ and F_1 is the set of all functions from $\mathbb{N}$ to $\{0, 1\}$.

Lemma 8.5 *The information system U'_1 belongs to the class $\mathscr{V}_1$.*

Proof It is easy to show that the information system U'_1 has infinite I-dimension. Therefore $U'_1 \notin \mathscr{D}$. We now show that $U'_1 \notin \mathscr{I}$. For a natural n, we define functions $f_0, f_1, \ldots, f_n \in F_1$. For any $a \in \mathbb{N}$, $f_0(a) = 1$ if and only if $a \in \{1, \ldots, n\}$. For $i = 1, \ldots, n$, $f_i(a) = 1$ if and only if $a = i$. It is easy to show that the equation system $\{f_0(x) = 1, f_1(x) = 0, \ldots, f_n(x) = 0\}$ is inconsistent but each proper subsystem of this system if consistent. Therefore $U'_1 \notin \mathscr{I}$. Using Lemma 8.4, we obtain $eind(U'_1) = (0, 0, 0, 0)$, i.e., $U'_1 \in \mathscr{V}_1$. □

Define an infinite binary information system $U_2' = (A_2, F_2)$ as follows: A_2 is the set of all infinite sequences of the kind $a_1, a_2, \ldots$, where $a_j \in \{0, 1\}$, $j \in \mathbb{N}$, $F_2 = \{f_i : i \in \mathbb{N}\}$ and $f_i(a_1, a_2, \ldots) = a_i$.

Lemma 8.6 *The information system U_2' belongs to the class $\mathscr{V}_2$.*

Proof It is easy to show that the information system U_2' has infinite I-dimension. Therefore $U_2' \notin \mathscr{D}$. Let S be a system of equations over U_2'. It is clear that S is inconsistent if and only if, for some $i \in \mathbb{N}$, the system S contains equations $f_i(x) = 0$ and $f_i(x) = 1$. Therefore U_2' is 2-i-restricted and $U_2' \in \mathscr{I}$. Using Lemma 8.4, we obtain $eind(U_2') = (0, 0, 0, 1)$, i.e., $U_2' \in \mathscr{V}_2$. □

For any $i \in \mathbb{N}$, we define two functions $p_i : \mathbb{N} \to \{0, 1\}$ and $l_i : \mathbb{N} \to \{0, 1\}$. Let $j \in \mathbb{N}$. Then $p_i(j) = 1$ if and only if $j = i$, and $l_i(j) = 1$ if and only if $j > i$.

Define an infinite binary information system $U_3' = (A_3, F_3)$ as follows: $A_3 = \mathbb{N}$ and $F_3 = \{p_i : i \in \mathbb{N}\} \cup \{l_i : i \in \mathbb{N}\}$.

Lemma 8.7 *The information system U_3' belongs to the class $\mathscr{V}_3$.*

Proof For $n \in \mathbb{N}$, denote $S_n = \{p_1(x) = 0, \ldots, p_n(x) = 0, l_n(x) = 0\}$. It is easy to show that the equation system S_n is inconsistent and each proper subsystem of S_n is consistent. Therefore $U_3' \notin \mathscr{I}$. By Lemma 8.3, $U_3' \notin \mathscr{R}$. Using attributes from the set $\{l_i : i \in \mathbb{N}\}$, we can construct d-complete tree over U_3' for each $d \in \mathbb{N}$. By Lemma 7.1 and Theorem 7.2, $U_3' \notin \mathscr{C}$. One can show that $I(U_3') = 1$. Therefore $U_3' \in \mathscr{D}$. Thus, $eind(U_3') = (0, 1, 0, 0)$, i.e., $U_3' \in \mathscr{V}_3$. □

Define an infinite binary information system $U_4' = (A_4, F_4)$ as follows: $A_4 = \mathbb{N}^2$ and $F_4 = \{f_i : i \in \mathbb{N}\} \cup \{f_{i,j} : i, j \in \mathbb{N}\}$. For any $a = (p, q) \in A_4$ and any $f_i \in F_4$, $f_i(a) = 1$ if and only if $p > i$. For any $a \in A_4$ and any $f_{i,j} \in F_4$, $f_{i,j}(a) = 1$ if and only if $a = (i, j)$.

Lemma 8.8 *The information system U_4' belongs to the class $\mathscr{V}_4$.*

Proof Let $n \in \mathbb{N}$ and $S_n = \{f_{1,1}(x) = 0, \ldots, f_{1,n}(x) = 0\}$. It is easy to show that the system S_n is consistent and each proper subsystem of S_n has another set of solutions on A_4 than the system S_n. Therefore $U_4' \notin \mathscr{R}$.

Using attributes from the set $\{f_i : i \in \mathbb{N}\}$, we can construct d-complete tree over U_4' for each $d \in \mathbb{N}$. By Lemma 7.1 and Theorem 7.2, $U_4' \notin \mathscr{C}$.

Let S be an equation system over U_4'. One can show that S is inconsistent if and only if S contains at least one of the following pairs of equations:

- $f_{i,j}(x) = 0$ and $f_{i,j}(x) = 1$;
- $f_{i,j}(x) = 1$ and $f_{k,l}(x) = 1$, $(i, j) \neq (k, l)$;

- $f_{i,j}(x) = 1$ and $f_k(x) = 0$, $i > k$;
- $f_{i,j}(x) = 1$ and $f_k(x) = 1$, $i \leq k$;
- $f_i(x) = 0$ and $f_j(x) = 1$, $i \leq j$.

Therefore U_4' is 2-i-restricted and $U_4' \in \mathscr{I}$. One can show that $I(U_4') = 1$. Therefore $U_4' \in \mathscr{D}$. Thus, $eind(U_4') = (0, 1, 0, 1)$, i.e., $U_4' \in \mathscr{V}_4$. □

Define an infinite binary information system $U_5' = (A_5, F_5)$ as follows: $A_5 = \bigcup_{i \in \mathbb{N}}\{(i, 1), \ldots, (i, i)\}$ and $F_5 = \bigcup_{i \in \mathbb{N}}\{f_i, f_{i,1}, \ldots, f_{i,i}\}$. For any $a \in A_5$ and any $f_i \in F_5$, $f_i(a) = 1$ if and only if $a \in \{(i, 1), \ldots, (i, i)\}$. For any $a \in A_5$ and any $f_{i,j} \in F_5$, $f_{i,j}(a) = 1$ if and only if $a = (i, j)$.

Lemma 8.9 *The information system U_5' belongs to the class $\mathscr{V}_5$.*

Proof It is easy to show that U_5' is 2-information system. In particular, $(A_5(f_i, 1), F_5)$ is 0-information system if $i = 1$, $(A_5(f_i, 1), F_5)$ is 1-information system if $i > 1$, and $(A_5(f_{i,j}, 1), F_5)$ is 0-information system for any attribute $f_{i,j} \in F_5$. Therefore $U_5' \in \mathscr{C}$. Let $i \in \mathbb{N}$ and $S_i = \{f_i(x) = 1, f_{i,1}(x) = 0, \ldots, f_{i,i}(x) = 0\}$. One can show that S_i is inconsistent and each proper subsystem of S_i is consistent. Therefore $U_5' \notin \mathscr{I}$. Using Lemma 8.4, we obtain $eind(U_3') = (0, 1, 1, 0)$, i.e., $U_5' \in \mathscr{V}_5$. □

Define an infinite binary information system $U_6' = (A_6, F_6)$ as follows: $A_6 = \mathbb{N}$ and $F_6 = \{p_i : i \in \mathbb{N}\}$.

Lemma 8.10 *The information system U_6' belongs to the class $\mathscr{V}_6$.*

Proof It is easy to show that U_6' is 1-information system: evidently, U_6' is not 0-information system, and $(A_6(p_i, 1), F_6)$ is 0-information system for any $i \in \mathbb{N}$. Therefore $U_6' \in \mathscr{C}$. Let S be an equation system over U_6'. One can show that S is inconsistent if and only if it contains equations $p_i(x) = 0$ and $p_i(x) = 1$ for some $i \in \mathbb{N}$ or it contains equations $p_i(x) = 1$ and $p_j(x) = 1$ for some $i, j \in \mathbb{N}$, $i \neq j$. Therefore U_6' is 2-i-restricted and $U_6' \in \mathscr{I}$. Using Lemma 8.4, we obtain $eind(U_6') = (0, 1, 1, 1)$, i.e., $U_6' \in \mathscr{V}_6$. □

Define an infinite binary information system $U_7' = (A_7, F_7)$ as follows: $A_7 = \mathbb{N}$ and $F_7 = \{l_i : i \in \mathbb{N}\}$.

Lemma 8.11 *The information system U_7' belongs to the class $\mathscr{V}_7$.*

Proof Let us consider an arbitrary consistent system of equations S over U_7'. We now show that there is a subsystem of S, which has at most two equations and the same set of solutions as S. Let S contain both equations of the kind $l_i(x) = 1$ and $l_j(x) = 0$. Denote $i_0 = \max\{i : l_i(x) = 1 \in S\}$ and $j_0 = \min\{j : l_j(x) = 0 \in S\}$. One can show that the system of equations $S' = \{l_{i_0}(x) = 1, l_{j_0}(x) = 0\}$ has the same set of solutions as S. The case when S contains for some $\delta \in \{0, 1\}$ only equations of the kind $l_p(x) = \delta$ can be considered in a similar way. In this case, the equation system S' contains only one equation. Therefore the information system U'_7 is 2-reduced and $U'_7 \in \mathscr{R}$. Using Lemma 8.4, we obtain $eind(U'_7) = (1, 1, 0, 1)$, i.e., $U'_7 \in \mathscr{V}_7$. □

Proof (of Theorem 8.2*)* From Lemma 8.4 it follows that, for any infinite binary information system, its extended indicator vector coincides with one of the rows of Table 8.1. Using Lemmas 8.5–8.11, we conclude that each row of Table 8.1 is the extended indicator vector of some infinite binary information system. □

8.5 Conclusions

Based on the results of exact learning, test theory, and rough set theory, for an arbitrary infinite binary information system, we considered five functions, which characterize the dependence in the worst case of the minimum depth of a decision tree solving a problem on the number of attributes in the problem description. These five functions correspond to (1) decision trees using attributes, (2) decision trees using arbitrary hypotheses, (3) decision trees using both attributes and arbitrary hypotheses, (4) decision trees using proper hypotheses, and (5) decision trees using both attributes and proper hypotheses. The first three functions were studied in Chap. 7. The last two functions were investigated in the present chapter: we described possible types of behavior for each of these two functions. We also studied joint behavior of all five functions and distinguished seven complexity classes of infinite binary information systems.

References

1. Angluin, D.: Queries revisited. Theor. Comput. Sci. **313**(2), 175–194 (2004)
2. Hegedüs, T.: Generalized teaching dimensions and the query complexity of learning. In: Maass W. (ed.) Eight Annual Conference on Computational Learning Theory, COLT 1995, Santa Cruz, CA, USA, 5–8 July 1995, pp. 108–117. ACM (1995)
3. Hellerstein, L., Pillaipakkamnatt, K., Raghavan, V., Wilkins, D.: How many queries are needed to learn? J. ACM **43**(5), 840–862 (1996)
4. Moshkov, M.: Test theory and problems of machine learning. In: International School-Seminar on Discrete Mathematics and Mathematical Cybernetics, Ratmino, Russia, 31 May–3 June 2001, pp. 6–10. MAX Press, Moscow (2001)
5. Moshkov, M.: Time complexity of decision trees. In: Peters, J.F., Skowron, A. (eds.) Trans. Rough Sets III, Lecture Notes in Computer Science, vol. 3400, pp. 244–459. Springer (2005)
6. Moshkov, M.: Exact learning and test theory. arXiv:2201.04506 [cs.CC] (2022). https://doi.org/10.48550/arXiv.2201.04506

7. Sauer, N.: On the density of families of sets. J. Comb. Theory (A) **13**, 145–147 (1972)
8. Shelah, S.: A combinatorial problem; stability and order for models and theories in infinitary languages. Pac. J. Math. **41**, 241–261 (1972)

9 Infinite Families of Concepts

In Chaps. 7 and 8, we investigated the time complexity of decision trees of five types over infinite binary information systems. The aim of the present chapter, which is based on [10], is to translate the obtained results into exact learning.

An exact learning problem is defined by a domain X and a concept class B [2]. The domain X is a nonempty finite set. A concept is any subset of X, and the concept class B is a nonempty set of concepts. For a given (but hidden) concept $c \in B$, we should recognize it using two types of queries. A membership query: for an element $x \in X$, we ask if $x \in c$. The answer is either $x \in c$ or $x \notin c$. An equivalence query: for a concept $c' \subseteq X$, we ask if $c' = c$. The answer is either yes or no. In the latter case, we receive a counterexample $x \in (c' \setminus c) \cup (c \setminus c')$, which is chosen nondeterministically. The concept c' can be considered as a hypothesis. This hypothesis and corresponding equivalence query are called proper if $c' \in B$.

To translate the results obtained in Chaps. 7 and 8 into exact learning, we introduce the notion of an infinite family of concepts consisting of an infinite set of elements U and an infinite set C of subsets of U called concepts. Each nonempty finite subset X of the set U describes a problem of exact learning with the domain X and the concept class $\{c \cap X : c \in C\}$. We study the following modification of this problem: for a given (but hidden) concept $c \in C$, we should recognize the concept $c \cap X$ using the same queries as for the initial problem. Both the problem and its modification have the same sets of decision trees solving them.

In this chapter, for each infinite family of concepts, we study five functions of Shannon type, which characterize the dependence in the worst case of the minimum depth of a decision tree solving an exact learning problem on the number of elements in the problem description. The considered five functions correspond to the following five types of decision trees:

1. Only membership queries are used in decision trees.
2. Only equivalence queries are used in decision trees.

M. Azad et al., *Decision Trees with Hypotheses*, Synthesis Lectures on Intelligent Technologies, https://doi.org/10.1007/978-3-031-08585-7_9

3. Both membership and equivalence queries are used in decision trees.
4. Only proper equivalence queries are used in decision trees.
5. Both membership and proper equivalence queries are used in decision trees.

The first function has two possible types of behavior—logarithmic and linear, and each of the remaining functions has three possible types of behavior—constant, logarithmic, and linear. We also consider joint behavior of these five functions and distinguish seven complexity classes of infinite families of concepts.

The translation of the results obtained in Chaps. 7 and 8 into exact learning will be useful for researchers in this area: the duality of the two theories is simple (attributes and elements from theory of decision trees with hypotheses correspond to elements and concepts from exact learning, respectively) but working out the details takes some effort.

The rest of the chapter is organized as follows. Sections 9.1 and 9.2 present the basic notions and main results. Sections 9.3–9.5 contain proofs, and Sect. 9.6—short conclusions.

9.1 Basic Notions

Let U be a nonempty set and C be a nonempty set of subsets of U called *concepts*. The pair $F = (U, C)$ is called a *family of concepts*. If U and C are infinite sets, then the pair $F = (U, C)$ is called an *infinite family of concepts*. For each element $u \in U$, we define a function $u : C \to \{0, 1\}$ with the same name as follows: for any $c \in C$, $u(c) = 1$ if and only if $u \in c$.

A *problem over* F is an arbitrary n-tuple $z = (u_1, \ldots, u_n)$, where $n \in \mathbb{N}$, $\mathbb{N}$ is the set of natural numbers $\{1, 2, \ldots\}$, and $u_1, \ldots, u_n \in U$. The problem z is as follows: for a given concept $c \in C$, we should recognize the tuple $z(c) = (u_1(c), \ldots, u_n(c))$. The number $\dim z = n$ is called the *dimension* of the problem z. Denote $U(z) = \{u_1, \ldots, u_n\}$ and $C(z) = \{c \cap U(z) : c \in C\}$. Note that the problem z can also be interpreted as the exact learning problem with the domain $U(z)$ and the concept class $C(z)$. We denote by $P(F)$ the set of problems over the family of concepts F.

A *system of equations over* F is an arbitrary equation system of the kind

$$\{v_1(x) = \delta_1, \ldots, v_m(x) = \delta_m\} ,$$

where $m \in \mathbb{N} \cup \{0\}$, $v_1, \ldots, v_m \in U$, and $\delta_1, \ldots, \delta_m \in \{0, 1\}$ (if $m = 0$, then the considered equation system is empty). This equation system is called a *system of equations over* z if $v_1, \ldots, v_m \in U(z)$. The considered equation system is called *consistent* (on C) if its set of solutions on C is nonempty. The set of solutions of the empty equation system coincides with C.

As algorithms for problem z solving, we consider decision trees with two types of *queries* —membership and equivalence. We can choose an element $u_i \in U(z)$ and ask about value of the function u_i. This *membership* query has two possible answers $\{u_i(x) = 0\}$ and $\{u_i(x) = 1\}$. We can formulate a *hypothesis over* z in the form $H = \{u_1(x) = \delta_1, \ldots, u_n(x) = \delta_n\}$, where $\delta_1, \ldots, \delta_n \in \{0, 1\}$, and ask about this hypothesis. This *equivalence* query has $n + 1$ possible answers: $H, \{u_1(x) = \neg\delta_1\}, \ldots, \{u_n(x) = \neg\delta_n\}$, where $\neg 1 = 0$ and $\neg 0 = 1$. The first answer means that the hypothesis is true. Other answers are counterexamples. This hypothesis and the corresponding equivalence query are called *proper* (for F) if the system of equations H is consistent on C.

A *decision tree over* z is a marked finite directed tree with the root in which

- Each *terminal* node is labeled with an n-tuple from the set $\{0, 1\}^n$.
- Each node, which is not terminal (such nodes are called *working*), is labeled with an element from the set $U(z)$ (a membership query) or with a hypothesis over z (an equivalence query).
- If a working node is labeled with an element u_i from $U(z)$, then there are two edges, which leave this node and are labeled with the systems of equations $\{u_i(x) = 0\}$ and $\{u_i(x) = 1\}$, respectively.
- If a working node is labeled with a hypothesis

  $$H = \{u_1(x) = \delta_1, \ldots, u_n(x) = \delta_n\}$$

  over z, then there are $n + 1$ edges, which leave this node and are labeled with the systems of equations $H, \{u_1(x) = \neg\delta_1\}, \ldots, \{u_n(x) = \neg\delta_n\}$, respectively.

Let Γ be a decision tree over z. A *complete path* in Γ is an arbitrary directed path from the root to a terminal node in Γ. We now define an equation system $\mathscr{S}(\xi)$ over F associated with the complete path ξ. If there are no working nodes in ξ, then $\mathscr{S}(\xi)$ is the empty system. Otherwise, $\mathscr{S}(\xi)$ is the union of equation systems assigned to the edges of the path ξ. We denote by $C(\xi)$ the set of solutions on C of the system of equations $\mathscr{S}(\xi)$ (if this system is empty, then its solution set is equal to C).

We will say that a decision tree Γ over z *solves the problem* z *relative to* F if, for each concept $c \in C$ and for each complete path ξ in Γ such that $c \in C(\xi)$, the terminal node of the path ξ is labeled with the tuple $z(c)$.

We now consider an equivalent definition of a decision tree solving a problem. Denote by $\Delta_F(z)$ the set of tuples $(\delta_1, \ldots, \delta_n) \in \{0, 1\}^n$ such that the system of equations $\{u_1(x) = \delta_1, \ldots, u_n(x) = \delta_n\}$ is consistent. The set $\Delta_F(z)$ is the set of all possible solutions to the problem z. Let $\Delta \subseteq \Delta_F(z)$, $u_{i_1}, \ldots, u_{i_m} \in \{u_1, \ldots, u_n\}$, and $\sigma_1, \ldots, \sigma_m \in \{0, 1\}$. Denote

$$\Delta(u_{i_1}, \sigma_1) \cdots (u_{i_m}, \sigma_m)$$

the set of all n-tuples $(\delta_1, \ldots, \delta_n) \in \Delta$ for which $\delta_{i_1} = \sigma_1, \ldots, \delta_{i_m} = \sigma_m$.

Let Γ be a decision tree over the problem z. We correspond to each complete path ξ in the tree Γ a word $\pi(\xi)$ in the alphabet $\{(u_i, \delta) : u_i \in U(z), \delta \in \{0, 1\}\}$. If the equation system

$\mathscr{S}(\xi)$ is empty, then $\pi(\xi)$ is the empty word. If $\mathscr{S}(\xi) = \{u_{i_1}(x) = \sigma_1, \ldots, u_{i_m}(x) = \sigma_m\}$, then $\pi(\xi) = (u_{i_1}, \sigma_1) \cdots (u_{i_m}, \sigma_m)$. The decision tree Γ over z solves the problem z relative to F if, for each complete path ξ in Γ, the set $\Delta_F(z)\pi(\xi)$ contains at most one tuple and if this set contains exactly one tuple, then the considered tuple is assigned to the terminal node of the path ξ.

As time complexity of a decision tree, we consider its *depth* that is the maximum number of working nodes in a complete path in the tree or, which is the same, the maximum length of a complete path in the tree. We denote by $h(\Gamma)$ the depth of a decision tree Γ.

Let $z \in P(F)$. We denote by $h_F^{(1)}(z)$ the minimum depth of a decision tree over z, which solves z relative to F and uses only membership queries. We denote by $h_F^{(2)}(z)$ the minimum depth of a decision tree over z, which solves z relative to F and uses only equivalence queries. We denote by $h_F^{(3)}(z)$ the minimum depth of a decision tree over z, which solves z relative to F and uses both membership and equivalence queries. We denote by $h_F^{(4)}(z)$ the minimum depth of a decision tree over z, which solves z relative to F and uses only proper equivalence queries. We denote by $h_F^{(5)}(z)$ the minimum depth of a decision tree over z, which solves z relative to F and uses both membership and proper equivalence queries.

For $i = 1, \ldots, 5$, we define a function of Shannon type $h_F^{(i)}(n)$ that characterizes dependence of $h_F^{(i)}(z)$ on $\dim z$ in the worst case. Let $i \in \{1, \ldots, 5\}$ and $n \in \mathbb{N}$. Then

$$h_F^{(i)}(n) = \max\{h_F^{(i)}(z) : z \in P(F), \dim z \leq n\} .$$

9.2 Main Results

Let $F = (U, C)$ be an infinite family of concepts and $r \in \mathbb{N}$. We will say that the family of concepts F is *r-reduced* if, for each consistent on C system of equations over F, there exists a subsystem of this system that has the same set of solutions on C and contains at most r equations. We denote by $\mathscr{R}'$ the class of infinite families of concepts each of which is r-reduced for some $r \in \mathbb{N}$.

The next theorem follows from the results obtained in [8], where we studied closed classes of test tables (decision tables). It also follows from the results obtained in [9], where we investigated the weighted depth of decision trees.

Theorem 9.1 *Let F be an infinite family of concepts. Then the following statements hold:*

(a) If $F \in \mathscr{R}'$, then $h_F^{(1)}(n) = \Theta(\log n)$.
(b) If $F \notin \mathscr{R}'$, then $h_F^{(1)}(n) = n$ for any $n \in \mathbb{N}$.

Let $F = (U, C)$ be an infinite family of concepts. A subset $\{u_1, \ldots, u_m\}$ of U is *shattered by C* if, for any $\delta_1, \ldots, \delta_m \in \{0, 1\}$, the system of equations $\{u_1(x) = \delta_1, \ldots, u_m(x) = \delta_m\}$ is consistent on the set C. This definition is equivalent to the standard one: a subset

$\{u_1, \ldots, u_m\}$ of U is shattered by C if, for any subset X of $\{u_1, \ldots, u_m\}$, there exists a concept $c \in C$ such that $X = \{u_1, \ldots, u_m\} \cap c$. The empty set of elements is shattered by C by definition. We now define the parameter $VC(F)$, which is called the *Vapnik-Chervonenkis dimension* or *VC-dimension* of the concept family F [13]. If, for each $m \in \mathbb{N}$, the set U contains a subset shattered by C of the cardinality m, then $VC(F) = \infty$. Otherwise, $VC(F)$ is the maximum cardinality of a subset of the set U shattered by C. We denote by $\mathscr{D}'$ the class of infinite families of concepts with finite VC-dimension.

Let $F = (U, C)$ be a family of concepts, which is not necessary infinite, $u \in U$, and $\delta \in \{0, 1\}$. Denote

$$C(u, \delta) = \{c : c \in C, u(c) = \delta\} .$$

We now define inductively the notion of *k-family of concepts*, $k \in \mathbb{N} \cup \{0\}$. The family of concepts F is called 0-family of concepts if all functions corresponding to elements from U are constant on the set C. Let, for some $k \in \mathbb{N} \cup \{0\}$, the notion of m-family of concepts be defined for $m = 0, \ldots, k$. The family of concepts F is called $(k + 1)$-family of concepts if it is not m-family of concepts for $m = 0, \ldots, k$ and, for any $u \in U$, there exist numbers $\delta \in \{0, 1\}$ and $m \in \{0, \ldots, k\}$ such that the family of concepts $(U, C(u, \delta))$ is m-family of concepts. It is easy to show by induction on k that if $F = (U, C)$ is k-family of concepts, then $F' = (U', C')$, where $U' \subseteq U$ and $C' = \{c \cap U' : c \in C\}$, is l-family of concepts for some $l \leq k$. We denote by $\mathscr{C}'$ the class of infinite families of concepts for each of which there exists $k \in \mathbb{N}$ such that the considered family is k-family of concepts.

From Lemma 9.6 below it follows that $\mathscr{C}' \subseteq \mathscr{D}'$. Therefore, for any infinite family of concepts F, either $F \in \mathscr{C}'$, or $F \in \mathscr{D}' \setminus \mathscr{C}'$, or $F \notin \mathscr{D}'$.

Theorem 9.2 *Let F be an infinite family of concepts. Then the following statements hold:*

(a) If $F \in \mathscr{C}'$*, then* $h_F^{(2)}(n) = O(1)$ *and* $h_F^{(3)}(n) = O(1)$.
(b) If $F \in \mathscr{D}' \setminus \mathscr{C}'$*, then* $h_F^{(2)}(n) = \Theta(\log n)$, $h_F^{(3)}(n) = \Omega(\frac{\log n}{\log \log n})$*, and* $h_F^{(3)}(n) = O(\log n)$.
(c) If $F \notin \mathscr{D}'$*, then* $h_F^{(2)}(n) = n$ *and* $h_F^{(3)}(n) = n$ *for any* $n \in \mathbb{N}$.

Let $F = (U, C)$ be an infinite family of concepts and $r \in \mathbb{N}$. We will say that the family of concepts F is *r-i-reduced* if, for each inconsistent on C system of equations over F, there exists a subsystem of this system that is inconsistent and contains at most r equations. We denote by $\mathscr{I}'$ the class of infinite families of concepts each of which is r-i-reduced for some $r \in \mathbb{N}$.

Since $\mathscr{C}' \subseteq \mathscr{D}'$, for any infinite family of concepts F, either $F \in \mathscr{C}' \cap \mathscr{I}'$, or $F \in (\mathscr{D}' \setminus \mathscr{C}') \cap \mathscr{I}'$, or $F \in \mathscr{D}' \setminus \mathscr{I}'$, or $F \notin \mathscr{D}'$.

Theorem 9.3 *Let F be an infinite family of concepts. Then the following statements hold:*

(a) If $F \in \mathscr{C}' \cap \mathscr{I}'$, then $h_F^{(4)}(n) = O(1)$ and $h_F^{(5)}(n) = O(1)$.
(b) If $F \in (\mathscr{D}' \setminus \mathscr{C}') \cap \mathscr{I}'$, then $h_F^{(4)}(n) = \Theta(\log n)$, $h_F^{(5)}(n) = \Omega(\frac{\log n}{\log\log n})$, and $h_F^{(5)}(n) = O(\log n)$.
(c) If $F \in \mathscr{D}' \setminus \mathscr{I}'$ and $i \in \{4, 5\}$, then $h_F^{(i)}(n) \geq n - 1$ for infinitely many $n \in \mathbb{N}$ and $h_F^{(i)}(n) \leq n$ for any $n \in \mathbb{N}$.
(d) If $F \notin \mathscr{D}'$, then $h_F^{(4)}(n) = n$ and $h_F^{(5)}(n) = n$ for any $n \in \mathbb{N}$.

Let F be an infinite family of concepts. We now consider the joint behavior of the functions $h_F^{(1)}(n), \ldots, h_F^{(5)}(n)$. It depends on the belonging of the family of concepts F to the classes $\mathscr{R}'$, $\mathscr{D}'$, $\mathscr{C}'$, and $\mathscr{I}'$. We correspond to the family of concepts F its *indicator vector* $ind(U) = (e_1, e_2, e_3, e_4) \in \{0, 1\}^4$ in which $e_1 = 1$ if and only if $F \in \mathscr{R}'$, $e_2 = 1$ if and only if $F \in \mathscr{D}'$, $e_3 = 1$ if and only if $F \in \mathscr{C}'$, and $e_4 = 1$ if and only if $F \in \mathscr{I}'$.

Theorem 9.4 *For any infinite family of concepts, its indicator vector coincides with one of the rows of Table 9.1. Each row of Table 9.1 is the indicator vector of some infinite family of concepts.*

For $i = 1, \ldots, 7$, we denote by $\mathscr{F}_i$ the class of all infinite families of concepts, which indicator vector coincides with the ith row of Table 9.1. Table 9.2 summarizes Theorems 9.1–9.4. The first column contains the name of *complexity class* $\mathscr{F}_i$. The next four columns describe the indicator vector of families of concepts from this class. The last five columns $h_F^{(1)}(n), \ldots, h_F^{(5)}(n)$ contain information about behavior of the functions $h_F^{(1)}(n), \ldots, h_F^{(5)}(n)$ for information systems from the class $\mathscr{F}_i$. The notation $\approx \log n$ in a column $h_F^{(i)}(n)$ means that $h_F^{(i)}(n) = \Omega(\frac{\log n}{\log\log n})$ and $h_F^{(i)}(n) = O(\log n)$. The notation $\approx n$ in a column $h_F^{(i)}(n)$ means that $h_F^{(i)}(n) \leq n$ for any $n \in \mathbb{N}$ and $h_F^{(i)}(n) \geq n - 1$ for infinitely many $n \in \mathbb{N}$.

Table 9.1 Possible indicator vectors of infinite families of concepts

	$\mathscr{R}'$	$\mathscr{D}'$	$\mathscr{C}'$	$\mathscr{I}'$
1	0	0	0	0
2	0	0	0	1
3	0	1	0	0
4	0	1	0	1
5	0	1	1	0
6	0	1	1	1
7	1	1	0	1

Table 9.2 Summary of Theorems 9.1–9.4

	$\mathscr{R}'$	$\mathscr{D}'$	$\mathscr{C}'$	$\mathscr{I}'$	$h_F^{(1)}(n)$	$h_F^{(2)}(n)$	$h_F^{(3)}(n)$	$h_F^{(4)}(n)$	$h_F^{(5)}(n)$
$\mathscr{F}_1$	0	0	0	0	n	n	n	n	n
$\mathscr{F}_2$	0	0	0	1	n	n	n	n	n
$\mathscr{F}_3$	0	1	0	0	n	$\Theta(\log n)$	$\approx \log n$	$\approx n$	$\approx n$
$\mathscr{F}_4$	0	1	0	1	n	$\Theta(\log n)$	$\approx \log n$	$\Theta(\log n)$	$\approx \log n$
$\mathscr{F}_5$	0	1	1	0	n	$O(1)$	$O(1)$	$\approx n$	$\approx n$
$\mathscr{F}_6$	0	1	1	1	n	$O(1)$	$O(1)$	$O(1)$	$O(1)$
$\mathscr{F}_7$	1	1	0	1	$\Theta(\log n)$	$\Theta(\log n)$	$\approx \log n$	$\Theta(\log n)$	$\approx \log n$

Note that it is possible to consider the union $\mathscr{F}_1 \cup \mathscr{F}_2$ of the complexity classes $\mathscr{F}_1$ and $\mathscr{F}_2$ as one complexity class since the functions $h_F^{(1)}(n), \ldots, h_F^{(5)}(n)$ have the same behavior for information systems from these classes. However, in this chapter, we study $\mathscr{F}_1$ and $\mathscr{F}_2$ as different complexity classes.

9.3 Proofs of Theorems 9.1 and 9.2

In this section, we prove Theorems 9.1 and 9.2.

***Proof (of Theorem** 9.1)* (a) Let $F = (U, C) \in \mathscr{R}'$. First, we prove that $h_F^{(1)}(n) = O(\log n)$. Since $F \in \mathscr{R}'$, there exists a natural r such that, for each consistent on C system of equations over F, there exists a subsystem of this system, which has the same set of solutions on C and contains at most r equations.

Let $z = (u_1, \ldots, u_m)$ be a problem over F. The number of equation systems over z containing at most r equations is at most the number of r-tuples of equations of the kind $u_i = \delta$, where $u_i \in \{u_1, \ldots, u_m\}$ and $\delta \in \{0, 1\}$. The latter number is equal to $(2m)^r$. Therefore $|\Delta_F(z)| \le (2m)^r$.

We consider a decision tree Γ, which solves the problem z relative to F and uses only membership queries. This tree is constructed by a halving algorithm that is similar to proposed in [7]. We will describe the work of Γ for an arbitrary concept c from C. This work consists of the following steps.

Denote $\Delta = \Delta_U(z)$. If $|\Delta| = 1$, then the only n-tuple from Δ is the solution $z(c)$ of the problem z for the concept c. Let $|\Delta| \ge 2$. For $i = 1, \ldots, m$, we denote by δ_i a number from $\{0, 1\}$ such that $|\Delta(u_i, \delta_i)| \ge |\Delta(u_i, \neg\delta_i)|$. Let k be the maximum number from $\{1, \ldots, m\}$ for which the equation system $\{u_1(x) = \delta_1, \ldots, u_k(x) = \delta_k\}$ is consistent on C. This system contains a subsystem $\{u_{i_1}(x) = \delta_{i_1}, \ldots, u_{i_t}(x) = \delta_{i_t}\}$, which has the same set of solutions on C and for which $t \le r$. We now show that $|\Delta(u_{i_1}, \delta_{i_1}) \cdots (u_{i_t}, \delta_{i_t})| \le |\Delta|/2$. If $k = m$, then $|\Delta(u_{i_1}, \delta_{i_1}) \cdots (u_{i_t}, \delta_{i_t})| \le 1 \le |\Delta|/2$. Let $k < m$. Then the equation system $\{u_{i_1}(x) = \delta_{i_1}, \ldots, u_{i_t}(x) = \delta_{i_t}, u_{k+1}(x) = \delta_{k+1}\}$ is inconsistent. Therefore $\Delta(u_{i_1}, \delta_{i_1}) \cdots (u_{i_t}, \delta_{i_t}) \subseteq$

$\Delta(u_{k+1}, \neg\delta_{k+1})$ and

$$|\Delta(u_{i_1}, \delta_{i_1}) \cdots (u_{i_t}, \delta_{i_t})| \le |\Delta|/2.$$

We sequentially compute values of the functions $u_{i_1}, \ldots, u_{i_t}$ for the concept c. If $u_{i_1}(c) = \delta_{i_1}, \ldots, u_{i_t}(c) = \delta_{i_t}$, then during the next step we will work with the set of tuples $\Delta' = \Delta(u_{i_1}, \delta_{i_1}) \cdots (u_{i_t}, \delta_{i_t})$ for which $|\Delta'| \le |\Delta|/2$. If, for some $p \in \{1, \ldots, t-1\}$, $u_{i_1}(c) = \delta_{i_1}, \ldots, u_{i_p}(c) = \delta_{i_p}$ and $u_{i_{p+1}}(c) = \neg\delta_{i_{p+1}}$, then during the next step we will work with the set of tuples

$$\Delta'' = \Delta(u_{i_1}, \delta_{i_1}) \cdots (u_{i_p}, \delta_{i_p})(u_{i_{p+1}}, \neg\delta_{i_{p+1}}).$$

It is clear that $|\Delta''| \le |\Delta|/2$. At this step, we compute values of at most r functions (make at most r membership queries) and reduce the number of m-tuples (possible solutions) by half.

Let during the work with the concept c, the decision tree Γ take q steps. After $(q-1)$th step, the number of remaining m-tuples will be at least two and at most $(2m)^r/2^{q-1}$. Therefore $2^q \le (2m)^r$ and $q \le r\log_2(2m)$. So during the processing of the concept c, the decision tree Γ computes values of at most $r^2 \log_2(2m)$ functions (makes at most $r^2 \log_2(2m)$ membership queries). Since c is an arbitrary concept from C, the depth of Γ is at most $r^2 \log_2(2m)$. Since z is an arbitrary problem over F, we obtain $h_F^{(1)}(n) = O(\log n)$.

We now show that $h_F^{(1)}(n) = \Omega(\log n)$. We prove by induction on n that, for any natural n, there is a problem $z_n = (u_1, \ldots, u_n)$ over F such that $|\Delta_F(z_n)| \ge n+1$. Since C is an infinite set of concepts, there exist two concepts $c_1, c_2 \in C$ and an element $u_1 \in U$ such that $u_1 \in c_1$ and $u_1 \notin c_2$. Therefore $|\Delta_F(z_1)| \ge 2$, where $z_1 = (u_1)$. Let, for some natural n, there exist a problem $z_n = (u_1, \ldots, u_n)$ over F such that $|\Delta_F(z_n)| \ge n+1$. Since C is an infinite set of concepts, there exist two concepts $c_3, c_4 \in C$ and an element $u_{n+1} \in U$ such that $\{u_1, \ldots, u_n\} \cap c_3 = \{u_1, \ldots, u_n\} \cap c_4$, $u_{n+1} \in c_3$, and $u_{n+1} \notin c_4$. One can show that $|\Delta_F(z_{n+1})| \ge n+2$, where $z_{n+1} = (u_1, \ldots, u_n, u_{n+1})$. Let $n \in \mathbb{N}$, $z_n = (u_1, \ldots, u_n)$ be a problem over F such that $|\Delta_F(z_n)| \ge n+1$, and Γ be a decision tree solving the problem z relative to F. Then Γ should have at least $n+1$ terminal nodes. One can show that the number of terminal nodes in the tree Γ is at most $2^{h(\Gamma)}$. Therefore $n+1 \le 2^{h(\Gamma)}$, $h(\Gamma) \ge \log_2(n+1)$, and $h_F^{(1)}(z) \ge \log_2(n+1)$. Thus, $h_F^{(1)}(n) = \Omega(\log n)$ and $h_F^{(1)}(n) = \Theta(\log n)$.

(b) Let $F = (U, C) \notin \mathscr{R}'$. One can show that, for any $n \in \mathbb{N}$, there is a consistent system of equations $S = \{u_1(x) = \delta_1, \ldots, u_n(x) = \delta_n\}$ over F for which each proper subsystem has different solution set than S. Let Γ be a decision tree solving the problem $z = (u_1, \ldots, u_n)$ relative to F. Then there is a complete path ξ in Γ such that $\Delta_F(z_n)\pi(\xi) = \{(\delta_1, \ldots, \delta_n)\}$. From here it follows that $S = \mathscr{S}(\xi)$. Therefore the complete path ξ contains at least n working nodes and $h(\Gamma) \ge n$. Taking into account that Γ is an arbitrary decision tree solving z relative to F, we obtain $h_F^{(1)}(z) \ge n$ and $h_F^{(1)}(n) \ge n$. It is easy to show that, for each problem z over U, the inequality $h_F^{(1)}(z) \le \dim z$ holds: to solve the problem z, it is enough to compute values of functions corresponding to all elements from the set $U(z)$. Thus, $h_F^{(1)}(n) = n$ for any $n \in \mathbb{N}$. □

We precede the proof of Theorem 9.2 by two lemmas.

Let $d \in \mathbb{N}$. A *d-complete tree over the family of concepts* $F = (U, C)$ is a marked finite directed tree with the root in which

- Each terminal node is not labeled.
- Each nonterminal node is labeled with an element $u \in U$. There are two edges leaving this node that are labeled with the systems of equations $\{u(x) = 0\}$ and $\{u(x) = 1\}$, respectively.
- The length of each complete path (path from the root to a terminal node) is equal to d.
- For each complete path ξ, the equation system $\mathcal{S}(\xi)$, which is the union of equation systems assigned to the edges of the path ξ, is consistent.

Let G be a d-complete tree over F and $U(G)$ be the set of all elements attached to the nonterminal nodes of the tree G. The number of nonterminal nodes in G is equal to $2^0 + 2^1 + \ldots + 2^{d-1} = 2^d - 1$. Therefore $|U(G)| \leq 2^d$.

The results mentioned in the following lemma are obtained by methods similar to used by Littlestone [5], Maass and Turán [6], and Angluin [2].

Lemma 9.1 *Let $F = (U, C)$ be a family of concepts, $d \in \mathbb{N}$, G be a d-complete tree over F, and z be a problem over U such that $U(G) \subseteq U(z)$. Then*

(a) $h_F^{(2)}(z) \geq d$.
(b) $h_F^{(3)}(z) \geq \frac{d}{\log_2(2d)}$.

Proof (a) We prove the inequality $h_F^{(2)}(z) \geq d$ by induction on d. Let $d = 1$. Then the tree G has the only one nonterminal node, which is labeled with an element u such that the function u is not constant on C. Therefore $|\Delta_F(z)| \geq 2$ and $h_F^{(2)}(z) \geq 1$. Let, for $t \in \mathbb{N}$ and for any natural d, $1 \leq d \leq t$, the considered statement hold. Assume now that $d = t + 1$, G is a d-complete tree over F, z is a problem over F such that $U(G) \subseteq U(z)$, and Γ is a decision tree over z with the minimum depth, which solves the problem z and uses only equivalence queries. Let u be the element attached to the root of the tree G and H be the hypothesis attached to the root of the decision tree Γ. Then there is an edge, which leaves the root of Γ and is labeled with the equation system $\{u(x) = \delta\}$, where the equation $u(x) = \neg\delta$ belongs to the hypothesis H. This edge enters to the root of the subtree of Γ, which will be denoted by Γ_u. There is an edge, which leaves the root of G and is labeled with the equation system $\{u(x) = \delta\}$. This edge enters to the root of the subtree of G, which will be denoted by G_δ. One can show that the decision tree Γ_u solves the problem z relative to the family of concepts $F' = (U, C(u, \delta))$ and G_δ is a t-complete tree over F'. It is clear that $U(G_\delta) \subseteq U(z)$. Using the inductive hypothesis, we obtain $h(\Gamma_u) \geq t$. Therefore $h(\Gamma) \geq t + 1 = d$ and $h_F^{(2)}(z) \geq d$.

(b) We now prove the inequality $h_F^{(3)}(z) \geq \frac{d}{\log_2(2d)}$. Let $z = (u_1, \ldots, u)$ and Γ be a decision tree over z with the minimum depth, which solves the problem z and uses both membership and equivalence queries. The d-complete tree G has 2^d complete paths $\xi_1, \ldots, \xi_{2^d}$. For $i = 1, \ldots, 2^d$, we denote by c_i a solution of the equation system $\mathscr{S}(\xi_i)$. Denote $B = \{c_1, \ldots, c_{2^d}\}$. We now show that the decision tree Γ contains a complete path, which length is at least $\frac{d}{\log_2(2d)}$. We describe the process of this path construction beginning with the root of Γ.

Let the root of Γ be labeled with an element u_{i_0}. For $\delta \in \{0, 1\}$, we denote by B^δ the set of solutions on B of the equation system $\{u_{i_0}(x) = \delta\}$ and choose $\sigma \in \{0, 1\}$ for which $|B^\sigma| = \max\{|B^0|, |B^1|\}$. It is clear that $|B^\sigma| \geq \frac{|B|}{2} \geq \frac{|B|}{2d}$. In the considered case, the beginning of the constructed path in Γ is the root of Γ, the edge that leaves the root and is labeled with the equation system $\{u_{i_0}(x) = \sigma\}$, and the node to which this edge enters.

Let as assume now that the root of Γ is labeled with a hypothesis $H = \{u_1(x) = \delta_1, \ldots, u_n(x) = \delta_n\}$. We denote by ξ_H the complete path in G for which the system of equations $\mathscr{S}(\xi_H)$ is a subsystem of H. Let the nonterminal nodes of the complete path ξ_H be labeled with the elements $u_{i_1}, \ldots, u_{i_d}$. For $j = 1, \ldots, d$, we denote by B_j the set of solutions on B of the equation system $\{u_{i_j}(x) = \neg\delta_{i_j}\}$. It is clear that $|B_1| + \cdots + |B_d| \geq |B| - 1$. Therefore there exists $l \in \{1, \ldots, d\}$ such that $|B_l| \geq \frac{|B|-1}{d} \geq \frac{|B|}{2d}$. In the considered case, the beginning of the constructed path in Γ is the root of Γ, the edge that leaves the root and is labeled with the equation system $\{u_{i_l}(x) = \neg\delta_{i_l}\}$, and the node to which this edge enters.

We continue the construction of the complete path in Γ in the same way such that after the tth query we will have at least $\frac{|B|}{(2d)^t}$ elements from B. The process of path construction will continue at least until $\frac{|B|}{(2d)^t} \leq 1$, i.e., at least until $\log_2 |B| \leq t \log_2(2d)$. Since $|B| = 2^d$, we have $h(\Gamma) \geq t \geq \frac{d}{\log_2(2d)}$ and $h_F^{(3)}(z) \geq \frac{d}{\log_2(2d)}$. □

Lemma 9.2 *Let $F = (U, C)$ be a family of concepts, $k \in \mathbb{N} \cup \{0\}$, and F be not m-family of concepts for $m = 0, \ldots, k$. Then there exists a $(k + 1)$-complete tree over F.*

Proof We prove the considered statement by induction on k. Let $k = 0$. In this case, F is not 0-family of concepts. Then there exists an element $u \in U$ for which the function u is not constant on C. Using this element, it is easy to construct 1-complete tree over F.

Let the considered statement hold for some k, $k \geq 0$. We now show that it also holds for $k + 1$. Let $F = (U, C)$ be a family of concepts, which is not m-family of concepts for $m = 1, \ldots, k + 1$. Then there exists an element $u \in U$ such that, for any $\delta \in \{0, 1\}$, the information system $F_\delta = (U, C(u, \delta))$ is not m-information system for $m = 1, \ldots, k$. Using the inductive hypothesis, we conclude that, for any $\delta \in \{0, 1\}$, there exists a $(k + 1)$-complete tree G_δ over F_δ. Denote by G a directed tree with root in which the root is labeled with the element u and, for any $\delta \in \{0, 1\}$, there is an edge that leaves the root, is labeled with the equation system $\{u(x) = \delta\}$, and enters the root of the tree G_δ. One can show that the tree G is a $(k + 2)$-complete tree over F. □

***Proof (of Theorem* 9.2)** It is clear that $h_F^{(3)}(z) \le h_F^{(2)}(z)$ for any problem z over F. Therefore $h_F^{(3)}(n) \le h_F^{(2)}(n)$ for any $n \in \mathbb{N}$.

(a) Let $k \in \mathbb{N} \cup \{0\}$. We now show by induction on k that, for each k-family of concepts F (not necessary infinite) for each problem z over F, the inequality $h_F^{(2)}(z) \le k$ holds. Let $F = (U, C)$ be a 0-family of concepts and z be a problem over F. Since all functions corresponding to elements from $U(z)$ are constant on C, the set $\Delta_F(z)$ contains only one tuple. Therefore the decision tree containing only one node labeled with this tuple solves the problem z relative to F, and $h_F^{(2)}(z) = 0$.

Let $k \in \mathbb{N} \cup \{0\}$ and, for each m, $0 \le m \le k$, the considered statement hold. Let us show that it holds for $k+1$. Let $F = (U, C)$ be a $(k+1)$-family of concepts and $z = (u_1, \dots, u_n)$ be a problem over F. For $i = 1, \dots, n$, choose a number $\delta_i \in \{0, 1\}$ such that the family of concepts $(U, C(u_i, \neg\delta_i))$ is m_i-family of concepts, where $1 \le m_i \le k$. Using the inductive hypothesis, we conclude that, for $i = 1, \dots, n$, there is a decision tree Γ_i over z, which uses only equivalence queries, solves the problem z over $(U, C(u_i, \neg\delta_i))$, and has depth at most m_i. We denote by Γ a decision tree in which the root is labeled with the hypothesis $H = \{u_1(x) = \delta_1, \dots, u_n(x) = \delta_n\}$, the edge leaving the root and labeled with H enters the terminal node labeled with the tuple $(\delta_1, \dots, \delta_n)$, and for $i = 1, \dots, n$, the edge leaving the root and labeled with $\{u_i(x) = \neg\delta_i\}$ enters the root of the tree Γ_i. One can show that Γ solves the problem z relative to F and $h(\Gamma) \le k+1$. Therefore, $h_F^{(2)}(z) \le k+1$ for any problem z over F.

Let $F \in \mathscr{C}'$. Then F is k-family of concepts for some natural k and, for each problem z over F, we have $h_F^{(3)}(z) \le h_F^{(2)}(z) \le k$. Therefore $h_F^{(2)}(n) = O(1)$ and $h_F^{(3)}(n) = O(1)$.

(b) Let $F = (U, C) \in \mathscr{D}' \setminus \mathscr{C}'$. First, we show that $h_F^{(2)}(n) = O(\log n)$. Let $z = (u_1, \dots, u_n)$ be an arbitrary problem over F. From Lemma 5.1 [9] it follows that $|\Delta_F(z)| \le (4n)^{VC(F)}$. The proof of this lemma is based on the results similar to ones obtained by Sauer [11] and Shelah [12]. We consider a decision tree Γ over z, which solves z relative to F and uses only equivalence queries. This tree is constructed by the halving algorithm [1, 5]. We describe the work of this tree for an arbitrary concept c from C. Set $\Delta = \Delta_F(z)$. If $|\Delta| = 1$, then the only n-tuple from Δ is the solution $z(c)$ of the problem z for the concept c. Let $|\Delta| \ge 2$. For $i = 1, \dots, m$, we denote by δ_i a number from $\{0, 1\}$ such that $|\Delta(u_i, \delta_i)| \ge |\Delta(u_i, \neg\delta_i)|$. The root of Γ is labeled with the hypothesis $H = \{u_1(x) = \delta_1, \dots, u_n(x) = \delta_n\}$. After this query either the problem z is solved (if the answer is H) or we halve the number of tuples in the set Δ (if the answer is a counterexample $\{u_i(x) = \neg\delta_i\}$). In the latter case, set $\Delta = \Delta_F(z)(u_i, \neg\delta_i)$. The decision tree Γ continues to work with the concept c and the set of n-tuples Δ in the same way. Let during the work with the concept c, the considered decision tree make q queries. After the $(q-1)$th query, the number of remaining n-tuples in the set Δ is at least two and at most $(4n)^{VC(F)}/2^{q-1}$. Therefore $2^q \le (4n)^{VC(F)}$ and $q \le VC(F)\log_2(4n)$. So during the processing of the concept c, the decision tree Γ makes at most $VC(F)\log_2(4n)$ queries. Since c is an arbitrary concept from C, the depth of Γ is at most $VC(F)\log_2(4n)$. Since z is an arbitrary problem over F, we obtain $h_F^{(2)}(n) = O(\log n)$. Therefore $h_F^{(3)}(n) = O(\log n)$.

Using Lemma 9.2 and the relation $F \notin \mathscr{C}'$, we obtain that, for any $d \in \mathbb{N}$, there exists d-complete tree G_d over F. Let $U(G_d) = \{u_1, \ldots, u_{n_d}\}$. We know that $n_d \leq 2^d$. Denote $z_d = (u_1, \ldots, u_{n_d})$. From Lemma 9.1 it follows that $h_F^{(2)}(z_d) \geq d$ and $h_F^{(3)}(z_d) \geq \frac{d}{\log_2(2d)}$. As a result, we have $h_F^{(2)}(2^d) \geq d$ and $h_F^{(3)}(2^d) \geq \frac{d}{\log_2(2d)}$. Let $n \in \mathbb{N}$ and $n \geq 8$. Then there exists $d \in \mathbb{N}$ such that $2^d \leq n < 2^{d+1}$. We have $d > \log_2 n - 1$, $h_F^{(2)}(n) \geq \log_2 n - 1$, $h_F^{(2)}(n) = \Omega(\log n)$, and $h_F^{(2)}(n) = \Theta(\log n)$. It is easy to show that the function $\frac{x}{\log_2(2x)}$ is nondecreasing for $x \geq 2$. Therefore $h_F^{(3)}(n) \geq \frac{\log_2 n - 1}{\log_2(2(\log_2 n - 1))}$ and $h_F^{(3)}(n) = \Omega(\frac{\log n}{\log \log n})$.

(c) Let $F = (U, C) \notin \mathscr{D}'$. We now consider an arbitrary problem $z = (u_1, \ldots, u_n)$ over F and a decision tree over z, which uses only equivalence queries and solves the problem z over F in the following way. For a given concept $c \in C$, the first query is about the hypothesis $H_1 = \{u_1(x) = 1, \ldots, u_n(x) = 1\}$. If the answer is H_1, then the problem z is solved for the concept c. If, for some $i \in \{1, \ldots, n\}$, the answer is $\{u_i(x) = 0\}$, then the second query is about the hypothesis H_2 obtained from H_1 by replacing the equality $u_i(x) = 1$ with the equality $u_i(x) = 0$, etc. It is clear that after at most n equivalence queries the problem z for the concept c will be solved. Thus, $h_F^{(2)}(z) \leq n$ and $h_F^{(3)}(z) \leq n$. Since z is an arbitrary problem over F, we have $h_F^{(2)}(n) \leq n$ and $h_F^{(3)}(n) \leq n$ for any $n \in \mathbb{N}$.

Let $n \in \mathbb{N}$. Since $F \notin \mathscr{D}'$, there exist elements $u_1, \ldots, u_n \in U$ such that, for any $(\delta_1, \ldots, \delta_n) \in \{0, 1\}^n$, the equation system $\{u_1(x) = \delta_1, \ldots, u_n(x) = \delta_n\}$ is consistent on C. We now consider the problem $z = (u_1, \ldots, u_n)$ and an arbitrary decision tree Γ over z, which solves the problem z over F and uses both membership and equivalence queries. Let us show that $h(\Gamma) \geq n$. If $n = 1$, then the considered inequality holds since $|\Delta_F(z)| \geq 2$. Let $n \geq 2$. It is easy to show that an equation system over z is inconsistent if and only if it contains equations $u_i(x) = 0$ and $u_i(x) = 1$ for some $i \in \{1, \ldots, n\}$. For each node v of the decision tree Γ, we denote by S_v the union of systems of equations attached to edges in the path from the root of Γ to v. A node v of Γ will be called consistent if the equation system S_v is consistent.

We now construct a complete path ξ in the decision tree Γ, which nodes are consistent. We will start constructing the path from the root that is a consistent node. Let the path reach a consistent node v of Γ. If v is a terminal node, then the path ξ is constructed. Let v be a working node labeled with an element $u_i \in U(z)$. Then there exists $\delta \in \{0, 1\}$ for which the system of equations $S_v \cup \{u_i(x) = \delta\}$ is consistent. Then the path ξ will pass through the edge leaving v and labeled with the system of equations $\{u_i(x) = \delta\}$. Let v be labeled with a hypothesis $H = \{u_1(x) = \delta_1, \ldots, u_n(x) = \delta_n\}$. If there exists $i \in \{1, \ldots, n\}$ such that the system of equations $S_v \cup \{u_i(x) = \neg\delta\}$ is consistent, then the path ξ will pass through the edge leaving v and labeled with the system of equations $\{u_i(x) = \neg\delta\}$. Otherwise, $S_v = H$ and the path ξ will pass through the edge leaving v and labeled with the system of equations H.

Let all edges in the path ξ be labeled with systems of equations containing one equation each. Since all nodes of ξ are consistent, the equation system $\mathscr{S}(\xi)$ is consistent. We now show that $\mathscr{S}(\xi)$ contains at least n equations. Let us assume that this system contains less than n equations. Then the set $\Delta_F(z)\pi(\xi)$ contains more than one n-tuple, which is impossible. Therefore the length of the path ξ is at least n. Let there be edges in ξ, which are labeled with hypotheses, and the first edge in ξ labeled with a hypothesis H leaves the node v. Then $S_v = H$ and the length of ξ is at least n. Therefore $h(\Gamma) \geq n$, $h_F^{(3)}(z) \geq n$, and $h_F^{(2)}(z) \geq n$. As a result, we obtain $h_F^{(3)}(n) \geq n$ and $h_F^{(2)}(n) \geq n$. Thus, $h_F^{(2)}(n) = n$ and $h_F^{(3)}(n) = n$ for any $n \in \mathbb{N}$. □

9.4 Proof of Theorem 9.3

In this section, we prove Theorem 9.3. First, we consider several auxiliary statements.

Lemma 9.3 *Let $F = (U, C)$ be a family of concepts, z be a problem over F, and Γ_1 be a decision tree over z that solves the problem z relative to F and uses both membership and proper equivalence queries. Then there exists a decision tree Γ_2 over z that solves the problem z relative to F, uses only proper equivalence queries, and satisfies the inequality $h(\Gamma_2) \leq 2^{h(\Gamma_1)} - 1$.*

Proof We prove this statement by the induction on the depth $h(\Gamma_1)$ of the decision tree Γ_1. Let $h(\Gamma_1) = 0$. Then, as the decision tree Γ_2, we can take the decision tree Γ_1. It is clear that $h(\Gamma_2) = 2^{h(\Gamma_1)} - 1$. We now assume that the considered statement is true for any family of concepts, any problem over this family, and any decision tree over the considered problem that solves this problem, uses both membership and proper equivalence queries, and has depth at most k, $k \geq 0$.

Let $F = (U, C)$ be a family of concepts, $z = (u_1, \ldots, u_n)$ be a problem over F, and Γ_1 be a decision tree over z that solves the problem z relative to F, uses both membership and proper equivalence queries, and satisfies the condition $h(\Gamma_1) = k + 1$. We now show that there exists a decision tree Γ_2 over z, which solves the problem z relative to F, uses only proper equivalence queries, and which depth is at most $2^{k+1} - 1$.

Let the root of Γ_1 be labeled with a proper hypothesis

$$H = \{u_1(x) = \delta_1, \ldots, u_n(x) = \delta_n\} .$$

Then there are $n + 1$ edges, which leave the root, are labeled with the systems of equations $H, \{u_1(x) = \neg\delta_1\}, \ldots, \{u_n(x) = \neg\delta_n\}$, and enter the roots of subtrees $G_0, G_1, \ldots, G_n$ of the tree Γ_1, respectively. It is clear that, for $i = 1, \ldots, n$, G_i is a decision tree over z, which solves the problem z relative to the family of concepts $F_i = (U, C(u_i, \neg\delta_i))$, uses only membership queries and proper equivalence queries for F_i, and satisfies the inequality $h(G_i) \leq k$. Using the inductive hypothesis, we obtain that, for $i = 1, \ldots, n$, there exists a decision tree G_i' over z that solves the problem z relative to F_i, uses only proper equivalence queries for F_i, and satisfies the inequalities $h(G_i') \leq 2^{h(G_i)} - 1 \leq 2^k - 1$. Let G_0' be the decision tree, which contains only one node labeled with the tuple $(\delta_1, \ldots, \delta_n)$. We denote by Γ_2 the decision tree over z that is obtained from the decision tree Γ_1 by replacing the subtrees $G_0, G_1, \ldots, G_n$ with the subtrees $G'_0, G'_1, \ldots, G'_n$. It is easy to show that Γ_2 is a decision tree over z, which solves the problem z relative to F, uses only proper equivalence queries for F, and satisfies the inequalities $h(\Gamma_2) \le 2^k - 1 + 1 \le 2^{h(\Gamma_1)} - 1$.

Let the root of Γ_1 be labeled with an element u_i. Then there are two edges, which leave the root, are labeled with the equation systems $\{u_i(x) = 0\}$ and $\{u_i(x) = 1\}$, and enter the roots of subtrees T_0 and T_1 of the tree Γ_1, respectively. It is clear that, for $p = 0, 1$, T_p is a decision tree over z, which solves the problem z relative to the family of concepts $F_p = (U, C(u_i, p))$, uses only membership queries and proper equivalence queries for F_p, and satisfies the inequality $h(T_p) \le k$. Using the inductive hypothesis, we obtain that, for $p = 0, 1$, there exists a decision tree T'_p over z that solves the problem z relative to F_p, uses only proper equivalence queries for U_p, and satisfies the inequalities $h(T'_p) \le 2^{h(T_p)} - 1 \le 2^k - 1$. We denote by T the decision tree obtained from the decision tree T'_0 by replacing each terminal node of T'_0 with the decision tree T'_1.

Denote by Γ_2 the decision tree obtained from T by the following transformation of each complete path ξ in T. If $C(\xi) = \emptyset$, then we keep the path ξ untouched. Let $C(\xi) \neq \emptyset$, $\bar{\delta} = (\delta_1, \ldots, \delta_n)$ be the tuple that was attached to the terminal node of the tree T'_0 through which the path ξ passes, and $\bar{\sigma} = (\sigma_1, \ldots, \sigma_n)$ be the tuple attached to the terminal node of ξ. Since $C(\xi) \neq \emptyset$, at least one of the tuples $\bar{\delta}$ and $\bar{\sigma}$ belongs to the set $\Delta_F(z)$. Let, for the definiteness, $\bar{\delta} \in \Delta_F(z)$. Denote $H = \{u_1(x) = \delta_1, \ldots, u_n(x) = \delta_n\}$. We replace the terminal node of the path ξ with the working node labeled with the hypothesis H, which is proper for F. There are $n + 1$ edges that leave this node and are labeled with the systems of equations $H, \{u_1(x) = \neg\delta_1\}, \ldots, \{u_n(x) = \neg\delta_n\}$, respectively. The edge labeled with H enters to the terminal node labeled with the tuple $\bar{\delta}$. All other edges enter to terminal nodes labeled with the tuple $\bar{\sigma}$. One can show that Γ_2 is a decision tree over z that solves the problem z relative to F, uses only proper equivalence queries for U, and satisfies the relations $h(\Gamma_2) \le 2(2^k - 1) + 1 = 2^{h(\Gamma_1)} - 1$. □

Lemma 9.4 *Let $F = (U, C)$ be an infinite family of concepts. Then $h_F^{(3)}(n) \le h_F^{(5)}(n) \le h_F^{(4)}(n) \le n$ and $h_F^{(2)}(n) \le h_F^{(4)}(n)$ for any $n \in \mathbb{N}$.*

Proof It is clear, that $h_F^{(3)}(z) \le h_F^{(5)}(z) \le h_F^{(4)}(z)$ and $h_F^{(2)}(z) \le h_F^{(4)}(z)$ for any problem z over U. Therefore $h_F^{(3)}(n) \le h_F^{(5)}(n) \le h_F^{(4)}(n)$ and $h_F^{(2)}(n) \le h_F^{(4)}(n)$ for any $n \in \mathbb{N}$.

We now consider an arbitrary problem $z = (u_1, \ldots, u_n)$ over F and a decision tree over z, which uses only proper equivalence queries for F and solves the problem z relative to F in the following way. For a given concept $c \in C$, the first query is about an arbitrary proper hypothesis $H_1 = \{u_1(x) = \delta_1, \ldots, u_n(x) = \delta_n\}$ for F. If the answer is H_1, then the problem z is solved for the concept c. If, for some $i \in \{1, \ldots, n\}$, the answer is $\{u_i(x) = \neg\delta_i\}$, then the second query is about a proper hypothesis $H_2 = \{u_1(x) = \sigma_1, \ldots, u_n(x) = \sigma_n\}$ such that $\sigma_i = \neg\delta_i$. If the answer is H_2, then the problem z is solved for the concept c. If, for some $j \in \{1, \ldots, n\}$, the answer is $\{u_j(x) = \neg\sigma_j\}$, then the third query is about a proper hypothesis $H_3 = \{u_1(x) = \gamma_1, \ldots, u_n(x) = \gamma_n\}$ such that $\gamma_i = \neg\delta_i$ and $\gamma_j = \neg\sigma_j$, etc. It is clear that after at most n queries the problem z for the concept c will be solved. Thus, $h_F^{(4)}(z) \leq n$. Since z is an arbitrary problem over F, we have $h_F^{(4)}(n) \leq n$ for any $n \in \mathbb{N}$. □

***Proof (of Theorem** 9.3)* (a) Let $r \in \mathbb{N}$. We now show by induction on $k \in \mathbb{N} \cup \{0\}$ that, for each r-i-reduced k-family of concepts F (not necessary infinite) for each problem z over F, the inequality $h_F^{(5)}(z) \leq rk$ holds.

Let $F = (U, C)$ be a r-i-reduced 0-family of concepts and z be a problem over F. Since all functions corresponding to elements from $U(z)$ are constant on C, the set $\Delta_F(z)$ contains only one tuple. Therefore the decision tree consisting of one node labeled with this tuple solves the problem z relative to F, and $h_F^{(5)}(z) = 0$.

Let $k \in \mathbb{N} \cup \{0\}$ and, for each m, $0 \leq m \leq k$, the considered statement hold. Let us show that it holds for $k + 1$. Let $F = (U, C)$ be a r-i-reduced $(k + 1)$-family of concepts and $z = (u_1, \ldots, u_n)$ be a problem over F. For $i = 1, \ldots, n$, choose a number $\delta_i \in \{0, 1\}$ such that the family of concepts $(U, C(u_i, \neg\delta_i))$ is m_i-family of concepts, where $1 \leq m_i \leq k$. It is easy to show that $(U, C(u_i, \neg\delta_i))$ is r-i-reduced family of concepts. Using the inductive hypothesis, we conclude that, for $i = 1, \ldots, n$, there is a decision tree Γ_i over z, which uses both membership queries and proper equivalence queries for $(U, C(u_i, \neg\delta_i))$, solves the problem z relative to $(U, C(u_i, \neg\delta_i))$, and has depth at most rm_i.

Let the hypothesis $H = \{u_1(x) = \delta_1, \ldots, u_n(x) = \delta_n\}$ be proper for F. We denote by T_1 a decision tree in which the root is labeled with the hypothesis H, the edge leaving the root and labeled with H enters the terminal node labeled with the tuple $(\delta_1, \ldots, \delta_n)$, and for $i = 1, \ldots, n$, the edge leaving the root and labeled with $\{u_i(x) = \neg\delta_i\}$ enters the root of the tree Γ_i. One can show that T_1 is a decision tree over z, which uses both membership queries and proper equivalence queries for F, solves the problem z relative to F, and satisfies the inequalities $h(T_1) \leq rk + 1 \leq r(k + 1)$.

Let the hypothesis H be not proper for F. Then the equation system $\{u_1(x) = \delta_1, \ldots, u_n(x) = \delta_n\}$ is inconsistent on C, and there exists its subsystem $\{u_{i_1}(x) = \delta_{i_1}, \ldots, u_{i_t}(x) = \delta_{i_t}\}$, which is inconsistent on C and for which $t \leq r$. We denote by G a decision tree over z with 2^t terminal nodes in which each terminal node is labeled with n-tuple $(0, \ldots, 0)$, and each complete path contains t working nodes labeled with elements $u_{i_1}, \ldots, u_{i_t}$ starting from the root. We denote by T_2 a decision tree obtained from the decision tree G by transformation of each complete path ξ in G. Let $\{u_{i_1}(x) = \sigma_1\}, \ldots, \{u_{i_t}(x) = \sigma_t\}$ be equation systems attached to edges leaving the working nodes of ξ labeled with the elements $u_{i_1}, \ldots, u_{i_t}$, respectively. If $(\sigma_1, \ldots, \sigma_t) = (\delta_{i_1}, \ldots, \delta_{i_t})$, then we keep the path ξ untouched. Otherwise, let j be the minimum number from the set $\{1, \ldots, t\}$ such that $\sigma_j = \neg\delta_{i_j}$. In this case, we replace the terminal node of the path ξ with the root of the decision tree Γ_{i_j}. One can show that T_2 is a decision tree over z, which uses both membership and proper equivalence queries, solves the problem z relative to F, and satisfies the inequalities $h(T_2) \leq rk + t \leq r(k + 1)$. Therefore, $h_F^{(5)}(z) \leq r(k + 1)$ for any problem z over F.

Let $F \in \mathscr{C}' \cap \mathscr{I}'$. Then F is r-i-reduced k-family of concepts for some natural r and k, and $h_F^{(5)}(z) \leq rk$ for each problem z over F. From Lemma 9.3 it follows that $h_F^{(4)}(z) \leq 2^{rk} - 1$ for each problem z over F. Therefore $h_F^{(4)}(n) = O(1)$ and $h_F^{(5)}(n) = O(1)$.

(b) Let $F = (U, C) \in (\mathscr{D}' \setminus \mathscr{C}') \cap \mathscr{I}'$. By Lemma 9.4, $h_F^{(5)}(n) \geq h_F^{(3)}(n)$ and $h_F^{(4)}(n) \geq h_F^{(2)}(n)$ for any $n \in \mathbb{N}$. Using the fact that $U \in \mathscr{D}' \setminus \mathscr{C}'$ and Theorem 9.2, we obtain $h_F^{(2)}(n) = \Omega(\log n)$ and $h_F^{(3)}(n) = \Omega(\frac{\log n}{\log \log n})$. Therefore $h_F^{(4)}(n) = \Omega(\log n)$ and $h_F^{(5)}(n) = \Omega(\frac{\log n}{\log \log n})$.

Since the family of concepts F belongs to the set $\mathscr{D}'$, it has finite VC-dimension $VC(F)$. Since $F \in \mathscr{I}'$, the family of concepts F is r-i-reduced for some natural r. We assume that $r \geq 2$. We can do it because each t-i-reduced family of concepts, $t \in \mathbb{N}$, is $(t+1)$-i-reduced.

We now show that $h_F^{(4)}(n) = O(\log n)$. Let $z = (u_1, \dots, u_n)$ be an arbitrary problem over F. From Lemma 5.1 [9] it follows that $|\Delta_F(z)| \leq (4n)^{VC(F)}$.

We consider a decision tree Γ over z, which solves the problem z relative to F and uses only proper equivalence queries. This tree is constructed by a variant of the halving algorithm [2–4]. We describe the work of this tree for an arbitrary concept c from C. Set $\Delta = \Delta_F(z)$. If $|\Delta| = 1$, then the only n-tuple from Δ is the solution $z(c)$ to the problem z for the concept c. Let $|\Delta| \geq 2$. For $i = 1, \dots, n$, we denote by δ_i a number from $\{0, 1\}$ such that $|\Delta(u_i, \delta_i)| \geq |\Delta(u_i, \neg\delta_i)|$.

Let the system of equations $H = \{u_1(x) = \delta_1, \dots, u_n(x) = \delta_n\}$ be consistent on C. In this case, the root of Γ is labeled with the proper hypothesis H. After this query, either the problem z will be solved (if the answer is H) or the number of remaining tuples in Δ will be at most $|\Delta|/2$ (if the answer is a counterexample $\{u_i(x) = \neg\delta_i\}$).

Let the system of equations H be inconsistent on C. For any inconsistent subsystem B of H, there exists a subsystem D of B, which is inconsistent and contains at most r equations. Then the system D contains at least one equation $u_i(x) = \delta_i$ such that $|\Delta(u_i, \neg\delta_i)| \geq |\Delta|/r$. If we assume the contrary, we obtain that the system D is consistent, which is impossible. Let $u_i \in \{u_1, \dots, u_n\}$. The element u_i is called balanced if $|\Delta(u_i, \neg\delta_i)| \geq |\Delta|/r$, and unbalanced if $|\Delta(u_i, \neg\delta_i)| < |\Delta|/r$.

We denote by H_u the subsystem of H consisting of all equations $u_i(x) = \delta_i$ from H with unbalanced elements u_i. We now show that the system H_u is consistent. Let us assume the contrary. Then it will contain at least one equation for balanced element, which is impossible. Let b be a solution from C to the system H_u, and $u_1(b) = \sigma_1, \dots, u_n(b) = \sigma_n$. Then the system of equations $P = \{u_1(x) = \sigma_1, \dots, u_n(x) = \sigma_n\}$ is consistent on C.

In the considered case, the root of Γ is labeled with the proper hypothesis P. After this query, either the problem z will be solved (if the answer is P), or the number of remaining tuples in Δ will be less than $|\Delta|/r$ (if the answer is a counterexample $\{u_i(x) = \neg\sigma_i\}$ and u_i is an unbalanced element), or the number of remaining tuples in Δ will be at most $|\Delta|/2$ (if the answer is a counterexample $\{u_i(x) = \neg\sigma_i\}$, $\sigma_i = \delta_i$, and u_i is a balanced element), or the number of remaining tuples in Δ will be at most $|\Delta|(1 - 1/r)$ (if the answer is a counterexample $\{u_i(x) = \neg\sigma_i\}$, $\sigma_i = \neg\delta_i$, and u_i is a balanced element).

After the first query (H or P) of the decision tree Γ, either the problem z will be solved or the number of remaining tuples in Δ will be at most $|\Delta|(1 - 1/r)$. In the latter case when the answer is a counterexample of the kind $\{u_i(x) = \neg\gamma_i\}$ ($\gamma_i = \delta_i$ if the first query is H and $\gamma_i = \sigma_i$ if the first query is P) set $\Delta = \Delta_U(z)(u_i, \neg\gamma_i)$. It is easy to show that the family of concepts $(U, C(u_i, \neg\gamma_i))$ is also r-i-reduced. The decision tree Γ continues to work with the concept c and the set of n-tuples Δ in the same way.

Let during the work with the concept c, the decision tree Γ make q queries. After the $(q-1)$th query, the number of remaining n-tuples in the set Δ is at least two and at most $(4n)^{VC(F)}(1-1/r)^{q-1}$. Therefore $(1+1/(r-1))^q \leq (4n)^{VC(F)}$ and $q\ln(1+1/(r-1)) \leq VC(F)\ln(4n)$. Taking into account that $\ln(1+1/m) > 1/(m+1)$ for any natural m, we obtain $q < rVC(F)\ln(4n)$. So during the processing of the concept c, the decision tree Γ makes at most $rVC(F)\ln(4n)$ queries. Since c is an arbitrary concept from C, the depth of Γ is at most $rVC(F)\ln(4n)$ and $h_F^{(4)}(z) \leq rVC(F)\ln(4n)$. Since z is an arbitrary problem over F, we obtain $h_F^{(4)}(n) = O(\log n)$. By Lemma 9.4, $h_F^{(5)}(n) = O(\log n)$.

(c) Let $F = (U, C) \in \mathscr{D}' \setminus \mathscr{I}'$. From Lemma 9.4 it follows that $h_F^{(5)}(n) \leq h_F^{(4)}(n) \leq n$ for any $n \in \mathbb{N}$. We now show that, for any $m \in \mathbb{N}$, there exists a natural n such that $n \geq m$, $h_F^{(4)}(n) \geq n-1$, and $h_F^{(5)}(n) \geq n-1$.

Let $m \in \mathbb{N}$. Since $F \notin \mathscr{I}'$, there exists a system of equations P over U with $n \geq m$ equations such that P is inconsistent but each proper subsystem of P is consistent on C. Let, for the definiteness, $P = \{u_1(x) = 0, \dots, u_n(x) = 0\}$. Consider the problem $z = (u_1, \dots, u_n)$ over F. Then, for $i = 1, \dots, n$, the set $\Delta_F(z)$ contains n-tuple $\bar{\delta}_i = (0, \dots, 0, 1, 0, \dots, 0)$ in which all digits with the exception of the ith one are equal to 0.

Let Γ be a decision tree over z that solves the problem z relative to F and uses both membership and proper equivalence queries. We consider a complete path ξ in Γ in which each edge is labeled with an equation system of the kind $\{u_i(x) = 0\}$, where $u_i \in U(z)$. Such complete path exists since P is not a proper hypothesis. Let $\pi(\xi) = (u_{i_1}, 0) \cdots (u_{i_t}, 0)$ for some $u_{i_1}, \dots, u_{i_t} \in U(z)$. Since Γ solves the problem z, the set $\Delta_F(z)\pi(\xi)$ contains at most one tuple. If we assume that $t < n-1$, we obtain that $\Delta_F(z)\pi(\xi)$ contains at least two tuples. Therefore $t \geq n-1$ and $h(\Gamma) \geq n-1$. Thus, $h_F^{(5)}(z) \geq n-1$, $h_F^{(5)}(n) \geq n-1$ and, by Lemma 9.4, $h_F^{(4)}(n) \geq n-1$.

(d) Let $F \notin \mathscr{D}'$. From Lemma 9.4 it follows that $h_F^{(3)}(n) \leq h_F^{(5)}(n) \leq h_F^{(4)}(n) \leq n$ for any $n \in \mathbb{N}$. By Theorem 9.2, $h_F^{(3)}(n) = n$ for any $n \in \mathbb{N}$. Thus, $h_F^{(5)}(n) = h_F^{(4)}(n) = n$ for any $n \in \mathbb{N}$. □

9.5 Proof of Theorem 9.4

First, we prove several auxiliary statements.

Lemma 9.5 $\mathscr{R}' \subseteq \mathscr{D}'$.

Proof Let $F \in \mathscr{R}'$. By Theorem 9.1, $h_F^{(1)}(n) = \Theta(\log n)$. Let us assume that $F \notin \mathscr{D}'$. Then, for any $n \in \mathbb{N}$, there exists a problem $z = (u_1, \dots, u_n)$ over F such that $|\Delta_F(z)| = 2^n$. Let Γ be a decision tree over z, which solves the problem z relative to F and uses only membership queries. Then Γ should have at least 2^n terminal nodes. One can show that the number of terminal nodes in the tree Γ is at most $2^{h(\Gamma)}$. Then $2^n \le 2^{h(\Gamma)}$, $h(\Gamma) \ge n$, and $h_F^{(1)}(z) \ge n$. Therefore $h_F^{(1)}(n) \ge n$ for any $n \in \mathbb{N}$, which is impossible. Thus, $\mathscr{R}' \subseteq \mathscr{D}'$. □

Lemma 9.6 $\mathscr{C}' \subseteq \mathscr{D}'$.

Proof Let $F \in \mathscr{C}'$. By Theorem 9.2, $h_F^{(2)}(n) = O(1)$. Let us assume that $F \notin \mathscr{D}'$. Then, by Theorem 9.2, $h_F^{(2)}(n) = n$ for any $n \in \mathbb{N}$, which is impossible. Therefore $\mathscr{C}' \subseteq \mathscr{D}'$. □

Lemma 9.7 $\mathscr{R}' \cap \mathscr{C}' = \emptyset$.

Proof Assume the contrary: $\mathscr{R}' \cap \mathscr{C}' \neq \emptyset$ and $F = (U, C) \in \mathscr{R}' \cap \mathscr{C}'$. Let $r, k \in \mathbb{N}$ and F be r-reduced k-family of concepts. We now consider an arbitrary problem $z = (u_1, \dots, u_n)$ over F and describe a decision tree Γ over z, which uses only membership queries, solves the problem z over F, and has depth at most kr.

For $i = 1, \dots, n$, let δ_i be a number from $\{0, 1\}$ such that $(U, C(u_i, \neg\delta_i))$ is m_i-family of concepts with $0 \le m_i < k$. Let t be the maximum number from the set $\{1, \dots, n\}$ such that the system of equations $S = \{u_1(x) = \delta_1, \dots, u_t(x) = \delta_t\}$ is consistent. Then there exists a subsystem $\{u_{i_1}(x) = \delta_{i_1}, \dots, u_{i_p}(x) = \delta_{i_p}\}$ of the system S, which has the same set of solutions as S and for which $p \le r$. For a given $c \in C$, the decision tree Γ computes sequentially values $u_{i_1}(c), \dots, u_{i_p}(c)$.

If, for some $q \in \{1, \dots, p\}$, $u_{i_1}(c) = \delta_{i_1}, \dots, u_{i_{q-1}}(c) = \delta_{i_{q-1}}$, and $u_{i_q}(c) = \neg\delta_{i_q}$, then the decision tree Γ continues to work with the problem z and the family of concepts $F' = (U, C')$, where C' is the set of solutions on C of the equation system $\{u_{i_1}(x) = \delta_{i_1}, \dots, u_{i_{q-1}}(x) = \delta_{i_{q-1}}, u_{i_q}(x) = \neg\delta_{i_q}\}$. One can show that F' is l'-family of concepts for some $l' \le m_{i_q} < k$.

Let $u_{i_1}(c) = \delta_{i_1}, \dots, u_{i_p}(c) = \delta_{i_p}$. If $t = n$, then $(\delta_1, \dots, \delta_n)$ is the solution of the problem z for the considered concept c. Let $t < n$. Then the decision tree Γ continues to work with the problem z and the family of concepts $F'' = (U, C'')$, where C'' is the set of solutions on C of the equation system $\{u_{i_1}(x) = \delta_{i_1}, \dots, u_{i_p}(x) = \delta_{i_p}\}$. We know that the equation system $\{u_1(x) = \delta_1, \dots, u_t(x) = \delta_t, u_{t+1}(x) = \delta_{t+1}\}$ is inconsistent. Therefore the system $\{u_{i_1}(x) = \delta_{i_1}, \dots, u_{i_p}(x) = \delta_{i_p}, u_{t+1}(x) = \delta_{t+1}\}$ is inconsistent. Hence $C'' \subseteq C(u_{t+1}, \neg\delta_{t+1})$ and F'' is l''-family of concepts for some $l'' \le m_{t+1} < k$.

As a result, after at most r membership queries, we either solve the problem z or reduce the consideration of the problem z over k-family of concepts F to the consideration of the problem z over some l-family of concepts, where $l < k$. After at most rk membership queries, we solve the problem z since each problem over 0-family of concepts has exactly one possible solution. Therefore $h_F^{(1)}(z) \le rk$ and $h_F^{(1)}(n) = O(1)$. By Theorem 9.1, $h_F^{(1)}(n) = \Theta(\log n)$. The obtained contradiction shows that $\mathscr{R}' \cap \mathscr{C}' = \emptyset$. □

Lemma 9.8 $\mathscr{R}' \subseteq \mathscr{I}'$.

Proof Let $F = (U, C) \in \mathscr{R}'$. Then F is r-restricted for some natural r. We now show that F is $(r + 1)$-i-restricted. Let S be an arbitrary inconsistent on C equation system over F and S' be a subsystem of S with the maximum number of equations that is consistent.

Since F is r-restricted, the system S' has a subsystem S'' with at most r equations and the same set of solutions on C as the system S'. It is clear that there exists an equation $u(x) = \delta$ from S such that the system of equations $S' \cup \{u(x) = \delta\}$ is inconsistent. Then the subsystem $S'' \cup \{u(x) = \delta\}$ of S with at most $r + 1$ equations is inconsistent. Therefore F is $(r + 1)$-i-restricted and $F \in \mathscr{I}'$. □

Let F be an infinite family of concepts and its indicator vector $ind(F)$ be equal to (e_1, e_2, e_3, e_4). The vector (e_1, e_2, e_3) will be called the *restricted indicator vector* for the family of concepts F and will be denoted $rind(U)$. In this vector, $e_1 = 1$ if and only if $F \in \mathscr{R}'$, $e_2 = 1$ if and only if $F \in \mathscr{D}'$, and $e_3 = 1$ if and only if $F \in \mathscr{C}'$.

Lemma 9.9 *For any infinite family of concepts, its restricted indicator vector coincides with one of the rows of Table 9.3 with numbers 1–4.*

Proof Table 9.3 contains as rows all 3-tuples from the set $\{0, 1\}^3$. We now show that rows with numbers 5–8 cannot be restricted indicator vectors of infinite families of concepts. Assume the contrary: there is $i \in \{5, 6, 7, 8\}$ such that the row with the number i is the restricted indicator vector of an infinite family of concepts F. If $i = 5$, then $F \in \mathscr{R}'$ and $F \notin \mathscr{D}'$, but this is impossible since, by Lemma 9.5, $\mathscr{R}' \subseteq \mathscr{D}'$. If $i = 6$, then $F \in \mathscr{C}'$ and $F \notin \mathscr{D}'$, but this is impossible since, by Lemma 9.6, $\mathscr{C}' \subseteq \mathscr{D}'$. If $i = 7$, then $F \in \mathscr{R}'$ and $F \notin \mathscr{D}'$, but this is impossible since, by Lemma 9.5, $\mathscr{R}' \subseteq \mathscr{D}'$. If $i = 8$, then $F \in \mathscr{R}'$ and $F \in \mathscr{C}'$, but this is impossible since, by Lemma 9.7, $\mathscr{R}' \cap \mathscr{C}' = \emptyset$. Therefore, for any infinite family of concepts, its restricted indicator vector coincides with one of the rows of Table 9.3 with numbers 1–4. □

Lemma 9.10 *For any infinite family of concepts, its indicator vector coincides with one of the rows of Table 9.1.*

Proof Let F be an infinite family of concepts and $ind(F) = (e_1, e_2, e_3, e_4)$. Then $rind(F) = (e_1, e_2, e_3)$. By Lemma 9.9, (e_1, e_2, e_3) is one of the rows of Table 9.3 with numbers 1–4. Therefore, for each infinite family of concepts, its indicator vector is an extension of one of the rows of Table 9.3 with numbers 1–4: it can be obtained from the row by adding the fourth digit, which is equal to 0 or 1. Table 9.4 contains all extensions of rows of Table 9.3 with numbers 1–4. We now show that the row with number 8 cannot be the indicator vector of an infinite family of concepts. Assume the contrary: there is an infinite family of concepts F' such that $ind(F') = (1, 1, 0, 0)$. Then $F' \in \mathscr{R}'$ and $F' \notin \mathscr{I}'$, but this is impossible since, by Lemma 9.8, $\mathscr{R}' \subseteq \mathscr{I}'$. Therefore, for any infinite family of concepts, its indicator vector coincides with one of the rows of Table 9.4 with numbers 1–7. Thus, it coincides with one of the rows of Table 9.1. □

We now define seven infinite families of concepts $F_1, \ldots, F_7$ and prove that these families belong to the complexity classes $\mathscr{F}_1, \ldots, \mathscr{F}_7$, respectively.

Table 9.3 All 3-tuples from the set $\{0, 1\}^3$

	$\mathscr{R}'$	$\mathscr{D}'$	$\mathscr{C}'$
1	0	0	0
2	0	1	0
3	0	1	1
4	1	1	0
5	1	0	0
6	0	0	1
7	1	0	1
8	1	1	1

Table 9.4 All extensions of rows 1–4 of Table 9.3

	$\mathscr{R}'$	$\mathscr{D}'$	$\mathscr{C}'$	$\mathscr{I}'$
1	0	0	0	0
2	0	0	0	1
3	0	1	0	0
4	0	1	0	1
5	0	1	1	0
6	0	1	1	1
7	1	1	0	1
8	1	1	0	0

Define an infinite family of concepts $F_1 = (U_1, C_1)$ as follows: U_1 is the set of all infinite sequences $u = (u^{(i)})_{i \in \mathbb{N}}$, where $u^{(i)} \in \{0, 1\}$ for any $i \in \mathbb{N}$, $C_1 = \{c_i : i \in \mathbb{N}\}$ and, for any $u = (u^{(i)})_{i \in \mathbb{N}} \in U_1$ and $c_i \in C_1$, $u \in c_i$ if and only if $u^{(i)} = 1$.

Lemma 9.11 *The family of concepts F_1 belongs to the class $\mathscr{F}_1$.*

Proof It is easy to show that the family of concepts F_1 has infinite VC-dimension. Therefore $F_1 \notin \mathscr{D}'$. We now show that $F_1 \notin \mathscr{I}'$. Let $n \in \mathbb{N}$. We now define elements $u_0 = (u_0^{(i)})_{i \in \mathbb{N}}, u_1 = (u_1^{(i)})_{i \in \mathbb{N}}, \ldots, u_n = (u_n^{(i)})_{i \in \mathbb{N}} \in U_1$. For any $i \in \mathbb{N}$, $u_0^{(i)} = 1$ if and only if $i \in \{1, \ldots, n\}$. For $j = 1, \ldots, n$, $u_j^{(i)} = 1$ if and only if $j = i$. It is easy to show that the equation system $\{u_0(x) = 1, u_1(x) = 0, \ldots, u_n(x) = 0\}$ is inconsistent on C_1 but each proper subsystem of this system is consistent. Therefore $F_1 \notin \mathscr{I}'$. Using Lemma 9.10, we obtain $ind(F_1) = (0, 0, 0, 0)$, i.e., $F_1 \in \mathscr{F}_1$. □

Define an infinite family of concepts $F_2 = (U_2, C_2)$ as follows: $U_2 = \mathbb{N}$ and C_2 is the set of all subsets of the set $\mathbb{N}$.

Lemma 9.12 *The family of concepts F_2 belongs to the class $\mathscr{F}_2$.*

Proof It is easy to show that the family of concepts F_2 has infinite VC-dimension. Therefore $F_2 \notin \mathscr{D}'$. Let S be a system of equations over U_2. It is clear that S is inconsistent if and only if, for some $i \in \mathbb{N}$, the system S contains equations $i(x) = 0$ and $i(x) = 1$. Therefore F_2 is 2-i-restricted and $F_2 \in \mathscr{I}'$. Using Lemma 9.10, we obtain $ind(U_2) = (0, 0, 0, 1)$, i.e., $F_2 \in \mathscr{F}_2$. □

Define an infinite family of concepts $F_3 = (U_3, C_3)$ as follows: $U_3 = \{p_i : i \in \mathbb{N}\} \cup \{l_i : i \in \mathbb{N}\}$ and $C_3 = \{c_i : i \in \mathbb{N}\}$, where $c_1 = \{p_1\}$ and, for $i \geq 2$, $c_i = \{p_i, l_1, \ldots, l_{i-1}\}$.

Lemma 9.13 *The family of concepts F_3 belongs to the class $\mathscr{F}_3$.*

Proof For $n \in \mathbb{N}$, denote $S_n = \{p_1(x) = 0, \ldots, p_n(x) = 0, l_n(x) = 0\}$. It is easy to show that the equation system S_n is inconsistent on C_3 and each proper subsystem of S_n is consistent. Therefore $F_3 \notin \mathscr{I}'$. By Lemma 9.8, $F_3 \notin \mathscr{R}'$. Using elements from the set $\{l_i : i \in \mathbb{N}\}$, we can construct d-complete tree over F_3 for each $d \in \mathbb{N}$. By Lemma 9.1 and Theorem 9.2, $F_3 \notin \mathscr{C}'$. One can show that $VC(F_3) = 1$. Therefore $F_3 \in \mathscr{D}'$. Thus, $ind(U_3) = (0, 1, 0, 0)$, i.e., $F_3 \in \mathscr{F}_3$. □

Define an infinite family of concepts $F_4 = (U_4, C_4)$ as follows: $U_4 = \{u_i : i \in \mathbb{N}\} \cup \{u_{i,j} : i, j \in \mathbb{N}\}$ and $C_4 = \{c_{p,q} : p, q \in \mathbb{N}\}$. For any $u_i \in U_4$ and any $c_{p,q} \in C_4$, $u_i \in c_{p,q}$ if and only if $p > i$. For any $u_{i,j} \in U_4$ and any $c_{p,q} \in C_4$, $u_{i,j} \in c_{p,q}$ if and only if $(p, q) = (i, j)$.

Lemma 9.14 *The family of concepts F_4 belongs to the class $\mathscr{F}_4$.*

Proof Let $n \in \mathbb{N}$ and $S_n = \{u_{1,1}(x) = 0, \ldots, u_{1,n}(x) = 0\}$. It is easy to show that the system S_n is consistent and each proper subsystem of S_n has another set of solutions on C_4 than the system S_n. Therefore $F_4 \notin \mathscr{R}'$.

Using elements from the set $\{u_i : i \in \mathbb{N}\}$, we can construct d-complete tree over F_4 for each $d \in \mathbb{N}$. By Lemma 9.1 and Theorem 9.2, $F_4 \notin \mathscr{C}'$.

Let S be an equation system over F_4. One can show that S is inconsistent if and only if S contains at least one of the following pairs of equations:

- $u_{i,j}(x) = 0$ and $u_{i,j}(x) = 1$;
- $u_{i,j}(x) = 1$ and $u_{k,l}(x) = 1$, $(i, j) \neq (k, l)$;
- $u_{i,j}(x) = 1$ and $u_k(x) = 0$, $i > k$;
- $u_{i,j}(x) = 1$ and $u_k(x) = 1$, $i \leq k$;
- $u_i(x) = 0$ and $u_j(x) = 1$, $i \leq j$.

Therefore F_4 is 2-i-restricted and $F_4 \in \mathscr{I}'$. One can show that $VC(F_4) = 1$. Therefore $F_4 \in \mathscr{D}'$. Thus, $ind(F_4) = (0, 1, 0, 1)$, i.e., $F_4 \in \mathscr{F}_4$. □

Define an infinite family of concepts $F_5 = (U_5, C_5)$ as follows:

$$U_5 = \bigcup_{i \in \mathbb{N}} \{u_i, u_{i,1}, \dots, u_{i,i}\}$$

and $C_5 = \bigcup_{i \in \mathbb{N}} \{c_{i,1}, \dots, c_{i,i}\}$. For any $u_i \in U_5$ and any $c \in C_5$, $u_i \in c$ if and only if $c \in \{c_{i,1}, \dots, c_{i,i}\}$. For any $u_{i,j} \in U_5$ and any $c \in C_5$, $u_{i,j} \in c$ if and only if $c = c_{i,j}$.

Lemma 9.15 *The family of concepts F_5 belongs to the class $\mathscr{F}_5$.*

Proof It is easy to show that F_5 is 2-family of concepts. In particular, $(U_5, C_5(u_1, 1))$ is 0-family of concepts, $(U_5, C_5(u_i, 1))$ is 1-family of concepts if $i \geq 2$, and, for any element $u_{i,j} \in U_5$, $(U_5, C_5(u_{i,j}, 1))$ is 0-family of concepts. Therefore $F_5 \in \mathscr{C}'$. Let $i \in \mathbb{N}$ and $S_i = \{u_i(x) = 1, u_{i,1}(x) = 0, \dots, u_{i,i}(x) = 0\}$. One can show that the equation system S_i is inconsistent and each proper subsystem of S_i is consistent. Therefore $F_5 \notin \mathscr{I}'$. Using Lemma 9.10, we obtain $ind(F_5) = (0, 1, 1, 0)$, i.e., $F_5 \in \mathscr{F}_5$. □

Define an infinite family of concepts $F_6 = (U_6, C_6)$ as follows: $U_6 = \{p_i : i \in \mathbb{N}\}$ and $C_6 = \{c_i : i \in \mathbb{N}\}$, where $c_i = \{p_i\}$ for any $i \in \mathbb{N}$.

Lemma 9.16 *The family of concepts F_6 belongs to the class $\mathscr{F}_6$.*

Proof It is easy to show that F_6 is 1-family of concepts: evidently, F_6 is not 0-family of concepts, and $(U_6, C_6(p_i, 1))$ is 0-family of concepts for any $i \in \mathbb{N}$. Therefore $F_6 \in \mathscr{C}'$. Let S be an equation system over F_6. One can show that S is inconsistent if and only if it contains equations $p_i(x) = 0$ and $p_i(x) = 1$ for some $i \in \mathbb{N}$ or it contains equations $p_i(x) = 1$ and $p_j(x) = 1$ for some $i, j \in \mathbb{N}$, $i \neq j$. Therefore F_6 is 2-i-restricted and $F_6 \in \mathscr{I}'$. Using Lemma 9.10, we obtain $ind(F_6) = (0, 1, 1, 1)$, i.e., $F_6 \in \mathscr{F}_6$. □

Define an infinite family of concepts $F_7 = (U_7, C_7)$ as follows: $U_7 = \{l_i : i \in \mathbb{N}\}$ and $C_7 = \{c_i : i \in \mathbb{N}\}$, where $c_1 = \emptyset$ and, for $i \geq 2$, $c_i = \{l_1, \dots, l_{i-1}\}$.

Lemma 9.17 *The family of concepts F_7 belongs to the class $\mathscr{F}_7$.*

Proof Let us consider an arbitrary consistent system of equations S over F_7. We now show that there is a subsystem of S, which has at most two equations and the same set of solutions as S. Let S contain both equations of the kind $l_i(x) = 1$ and $l_j(x) = 0$. Denote $i_0 = \max\{i : l_i(x) = 1 \in S\}$ and $j_0 = \min\{j : l_j(x) = 0 \in S\}$. One can show that the system of equations $S' = \{l_{i_0}(x) = 1, l_{j_0}(x) = 0\}$ has the same set of solutions as S. The case when S contains for some $\delta \in \{0, 1\}$ only equations of the kind $l_p(x) = \delta$ can be considered in a similar way. In this case, the equation system S' contains only one equation. Therefore the family of concepts F_7 is 2-reduced and $F_7 \in \mathscr{R}'$. Using Lemma 9.10, we obtain $ind(F_7) = (1, 1, 0, 1)$, i.e., $F_7 \in \mathscr{F}_7$. □

***Proof (of Theorem** 9.4)* From Lemma 9.10 it follows that, for any infinite family of concepts, its indicator vector coincides with one of the rows of Table 9.1. Using Lemmas 9.11–9.17, we conclude that each row of Table 9.1 is the indicator vector of some infinite family of concepts. □

9.6 Conclusions

Based on the results of exact learning, test theory, and rough set theory, for an arbitrary infinite family of concepts, we studied five functions, which characterize the dependence in the worst case of the minimum depth of a decision tree solving a problem of exact learning on the number of elements in the problem description. These five functions correspond to (1) decision trees using membership queries, (2) decision trees using equivalence queries, (3) decision trees using both membership and equivalence queries, (4) decision trees using proper equivalence queries, and (5) decision trees using both membership and proper equivalence queries. We described possible types of behavior for each of these five functions. We also studied join behavior of these functions and distinguished seven complexity classes of infinite family of concepts.

References

1. Angluin, D.: Queries and concept learning. Mach. Learn. **2**(4), 319–342 (1988)
2. Angluin, D.: Queries revisited. Theor. Comput. Sci. **313**(2), 175–194 (2004)
3. Hegedüs, T.: Generalized teaching dimensions and the query complexity of learning. In: Maass, W. (ed.) Eight Annual Conference on Computational Learning Theory, COLT 1995, Santa Cruz, CA, USA, 5–8 July 1995, pp. 108–117. ACM (1995)
4. Hellerstein, L., Pillaipakkamnatt, K., Raghavan, V., Wilkins, D.: How many queries are needed to learn? J. ACM **43**(5), 840–862 (1996)
5. Littlestone, N.: Learning quickly when irrelevant attributes abound: a new linear-threshold algorithm. Mach. Learn. **2**(4), 285–318 (1988)
6. Maass, W., Turán, G.: Lower bound methods and separation results for on-line learning models. Mach. Learn. **9**, 107–145 (1992)
7. Moshkov, M.: Conditional tests. In: Yablonskii, S.V. (ed.) Problemy Kibernetiki, vol. 40, pp. 131–170. Nauka Publishers, Moscow (1983). (in Russian)
8. Moshkov, M.: On depth of conditional tests for tables from closed classes. In: Markov, A.A. (ed.) Combinatorial-Algebraic and Probabilistic Methods of Discrete Analysis, pp. 78–86. Gorky University Press, Gorky (1989). (in Russian)
9. Moshkov, M.: Time complexity of decision trees. In: Peters, J.F., Skowron, A. (eds.) Trans. Rough Sets III. Lecture Notes in Computer Science, vol. 3400, pp. 244–459. Springer (2005)

10. Moshkov, M.: Exact learning for infinite families of concepts. arXiv:2201.08225 [cs.AI] (2022). https://doi.org/10.48550/arXiv.2201.08225
11. Sauer, N.: On the density of families of sets. J. Comb. Theory (A) **13**, 145–147 (1972)
12. Shelah, S.: A combinatorial problem; stability and order for models and theories in infinitary languages. Pac. J. Math. **41**, 241–261 (1972)
13. Vapnik, V.N., Chervonenkis, A.Y.: On the uniform convergence of relative frequencies of events to their probabilities. Theory Probab. Appl. **16**(2), 264–280 (1971)

Appendix
Computation of Boolean Functions by Decision Trees with Hypotheses

In the appendix, we discuss some possibilities of using decision trees with hypotheses for computation of Boolean functions.

We can use conventional decision trees to compute a Boolean function

$$f(x_0, \ldots, x_n) .$$

In such a tree, each query is the recognition of the value of a variable x_i, $i \in \{0, \ldots, n\}$. Performing this operation directly takes one clock cycle. In decision trees with hypotheses, we can also use other queries: we can choose a tuple $(c_0, \ldots, c_n) \in \{0, 1\}^{n+1}$ and ask if the hypothesis $H = \{x_0 = c_0, \ldots, x_n = c_n\}$ is true. Either this hypothesis will be confirmed or we obtain a counterexample of the kind $x_i = \neg c_i$, where $i \in \{0, \ldots, n\}$. Performing this operation directly takes $n + 1$ clock cycles.

We assume now that $n = 63$ and describe a way to simulate the query based on the hypothesis H using a small number of clock cycles. First, we compare tuples $(x_0, \ldots, x_n)$ and $(c_0, \ldots, c_n)$. This operation usually requires one clock cycle. If the considered tuples are equal, then the hypothesis H is true. Otherwise, it is false and we need to return a counterexample. To this end, using the operation bitwise XOR, we compute the tuple $(b_0, \ldots, b_n) = (x_0 \oplus c_0, \ldots, x_n \oplus c_n)$, where $x_i \oplus c_i = 0$ if $x_i = c_i$ and $x_i \oplus c_i = 1$ if $x_i \neq c_i$. This operation usually requires one clock cycle. After that, we find the maximum number $i \in \{0, \ldots, n\}$ for which $b_i = 1$. We denote this number by i_0 and return the counterexample $x_{i_0} = \neg c_{i_0}$.

To find i_0, we correspond to the tuple $(b_0, \ldots, b_n)$ the unsigned integer

$$B = \sum_{i=0}^{n} b_i 2^i$$

M. Azad et al., *Decision Trees with Hypotheses*, Synthesis Lectures on Intelligent Technologies, https://doi.org/10.1007/978-3-031-08585-7

and compare it with numbers of the kind 2^j, $j \in \{0, \ldots, n\}$. It is clear that if $B \geq 2^j$, then $i_0 \geq j$ and if $B < 2^j$, then $i_0 < j$. The operation of comparison usually requires one clock cycle. We will use binary search. First, we compare B and 2^{32}. As a result, we localize i_0 either in the set $\{0, \ldots, 31\}$ or in the set $\{32, \ldots, 63\}$ each containing 32 numbers. After that we will work with one of these sets, etc. To find i_0 we will need 6 comparisons. In total, to simulate the considered query based on the hypothesis H, we will need 8 bitwise XOR and comparison operations, i.e., 8 clock cycles. Note that here and later we count only bitwise XOR and comparison operations and do not count time required by auxiliary operations with memory, etc.

We now describe what to do if $n \neq 63$. Let $t = \lceil \frac{n+1}{64} \rceil$. We divide tuples $(x_0, \ldots, x_n)$ and $(c_0, \ldots, c_n)$ into blocks of the length 64 starting from the beginning (if the last block in both tuples has the length less than 64, then we complement the last blocks with zeros to the length 64). For $j = 1, \ldots, t$, we compare the jth block in the extended tuple $(x_0, \ldots, x_n)$ and the jth block in the extended tuple $(c_0, \ldots, c_n)$. After t comparisons, we either confirm that the hypothesis H is true or find $j \in \{1, \ldots, t\}$ such that the jth block in the extended tuple $(x_0, \ldots, x_n)$ is different from the jth block in the extended tuple $(c_0, \ldots, c_n)$. In the latter case we need additionally one operation bitwise XOR and 6 comparisons to find a counterexample. In the worst case, we need $t + 7 = \lceil \frac{n+1}{64} \rceil + 7$ clock cycles. In particular, if $n + 1 = 1024$, then to simulate the query related to the hypothesis H, we need 22 comparisons and one operation bitwise XOR, i.e., 23 clock cycles.

Let us consider the problem of computation of conjunction of 1024 variables $x_0 \wedge \cdots \wedge x_{1023}$. If we simulate the work of a conventional decision tree computing this function, we need at least 1024 comparisons, i.e., at least 1024 clock cycles in the worst case. If we use the decision tree with hypotheses of the depth 1, which uses the hypothesis $\{x_0 = 1, \ldots, x_{1023} = 1\}$, we need 23 operations bitwise XOR and comparisons, i.e., 23 clock cycles in the worst case. Really, we need only 16 clock cycles since we only should check if the hypothesis is true and do not need to return counterexamples.

Similar approach can be used in the case, when the considered Boolean function f has n variables and exactly k, $k \ll n$, tuples of values of these variables on which its value is equal to 1 (we call them 1-tuples for f). We can build a conventional decision tree of the depth at most $k - 1$ satisfying the following condition. For each terminal node v of this tree, there is exactly one 1-tuple a_v such that the computation for this tuple finishes in this node. We will label each terminal node v of this tree with the hypothesis that checks if the tuple of values of the variables is equal to a_v. The output is equal to 1 if the considered tuples are equal and 0 otherwise. It is easy to see that the obtained decision tree computes the function f and has depth at most k. In the worst case, the simulation of the work of this tree requires $\lceil \frac{n}{64} \rceil + k + 6$ bitwise XOR and comparison operations. Really, we need only $\lceil \frac{n}{64} \rceil + k - 1$ clock cycles since it is not necessary to return counterexamples for the considered hypotheses.

We say that a 1-tuple is isolated if, for each tuple different from it in exactly one digit, the function f has value 0. One can show that if among the considered k 1-tuples there exists an isolated 1-tuple, then the minimum depth of a conventional decision tree computing f is equal to n.

Let us consider the Boolean function f with n variables, which is equal to 1 if and only if the sum of its variables is at least k, $1 \le k \le 63 \le n$. One can prove that the minimum depth of a conventional decision tree computing this function is equal to n.

We now show that there exists a decision tree with hypotheses that computes this function and has depth k. First, we ask about hypothesis in which each variable is equal to 0. If this hypothesis is confirmed, then the sum of variables is equal to 0 and the value of the function f is determined. Otherwise, we obtain a counterexample $x_{i_1} = 1$. If $k > 1$, then the second query is about a hypothesis in which $x_{i_1} = 1$ and all other variables are equal to 0. If this hypothesis is confirmed, then the sum of variables is equal to 1 and the value of the function f is determined. Otherwise, we obtain a counterexample $x_{i_2} = 1$. If $k > 2$, then the third query is about a hypothesis in which $x_{i_1} = 1, x_{i_2} = 1$, and all other variables are equal to 0, etc. It is clear that after at most k queries the value of the function f will be determined. In the worst case, the simulation of the work of this decision tree requires $k(\lceil \frac{n}{64} \rceil + 7)$ bitwise XOR and comparison operations.

In particular, if $n = 1024$ and $k = 10$, then to simulate the work of the considered decision tree with hypotheses, we need 230 operations of comparison and bitwise XOR.

Final Remarks

In this book, we studied decision trees that can use both queries based on one attribute and queries based on hypotheses about values of all attributes.

To investigate such decision trees, we created tools that, for a given decision table, can (i) find complexity of optimal decision trees and construct such trees using dynamic programming algorithms, (ii) construct decision trees using various greedy heuristics, and (iii) derive decision rules from the constructed decision trees and study their length and coverage. These tools can also be used for the study of exact learning problems.

To understand the difference between conventional decision trees and decision trees with hypotheses, we made computer experiments with 10 decision tables from the UCI ML Repository and 100 randomly generated Boolean functions with n variables, where $n = 3, \ldots, 6$.

For an arbitrary infinite binary information system, we investigated five functions of Shannon type corresponding to five types of decision trees, which characterize the dependence in the worst case of the minimum depth of a decision tree solving a problem on the number of attributes in the problem description. We studied joint behavior of these functions and described seven complexity classes of infinite binary information systems. The obtained results were translated into exact learning using the notion of an infinite family of concepts.

In the future, it would be interesting to extend the obtained results to the case of so-called partial hypotheses, which include only a part of the attributes from the considered decision table or from the problem under consideration.

M. Azad et al., *Decision Trees with Hypotheses*, Synthesis Lectures on Intelligent Technologies, https://doi.org/10.1007/978-3-031-08585-7

Index

M. Azad et al., *Decision Trees with Hypotheses*, Synthesis Lectures on Intelligent Technologies, https://doi.org/10.1007/978-3-031-08585-7

The manufacturer's authorised representative in the EU is Springer Nature Customer Service Centre GmbH, Europaplatz 3, 69115 Heidelberg, Germany. If you have any concerns regarding our products, please contact ProductSafety@springernature.com

Printed and bound by CPI Group (UK) Ltd, Croydon, CR0 4YY
28/09/2026
02224070-0020